Puzzle and Proof

Puzzle and Proof: A Decade of Problems from the Utah Math Olympiad is a compilation of the problems and solutions for the first 10 years of the Utah Math Olympiad. The problems are challenging but should be understandable at a high school level. Besides putting all problems in one place (70 in total), which have not previously appeared in print, the book provides additional inspiration for many of the problems and will contain the first published solutions for 10 problems that were originally published on the contest flyer. The book will be a fantastic resource for anyone who enjoys mathematical and/or logic puzzles or is interested in studying for mathematics competitions.

Features
- 70 carefully designed, high-quality high-school level math proof problems, with full solutions
- Detailed pictures and diagrams throughout to aid understanding
- Suitable for anyone with high school-level mathematics skills with an interest in furthering their understanding, or just enjoying the puzzles
- Solutions in the back of the book, sorting the problems by difficulty and topic.

Samuel Dittmer is a three-time USA Mathematical Olympiad qualifier and two-time William Lowell Putnam Mathematical Competition honorable mention from Indiana who has also lived in Albania. Sam won the national competition of the American Regions Mathematics League during his sophomore year in high school. Sam graduated with a PhD in math from UCLA in 2019. He lives with his wife and dog in Los Angeles.

Hiram Golze is a math teacher at the Waterford School. He grew up in Utah, where he won the Utah State Math Contest twice and qualified for the USA Mathematical Olympiad once. He currently coaches the Utah American Regions Mathematics League Team and has coached the Utah state ATHCOUNTS team. During the summers, he teaches at the AwesomeMath Summer Program. He has a master's degree in math from the University of Illinois at Urbana-Champaign.

Grant Molnar grew up in North Carolina. As an undergraduate, he tied for third place in the Virginia Tech Regional Math Competition and served as Brigham Young University Putnam Team Captain from 2014 to 2016. He graduated with his PhD in math from Dartmouth in 2023 and now works as a software developer.

Caleb Stanford is an assistant professor of computer science at UC Davis. He grew up in Utah, where he won the Utah State Math Contest five times and qualified for the USA Mathematical Olympiad once. He holds a bachelor's degree in math and computer science from Brown and a PhD in computer science from the University of Pennsylvania. He currently lives in the San Francisco Bay Area with his wife and two cats.

AK Peters/CRC Recreational Mathematics Series

Series Editors

Robert Fathauer
Snezana Lawrence
Jun Mitani
Colm Mulcahy
Peter Winkler
Carolyn Yackel

Mathematical Conundrums
Barry R. Clarke

Lateral Solutions to Mathematical Problems
Des MacHale

Basic Gambling Mathematics
The Numbers Behind the Neon, Second Edition
Mark Bollman

Design Techniques for Origami Tessellations
Yohei Yamamoto, Jun Mitani

Mathematicians Playing Games
Jon-Lark Kim

Electronic String Art
Rhythmic Mathematics
Steve Erfle

Playing with Infinity
Turtles, Patterns, and Pictures
Hans Zantema

Parabolic Problems
60 Years of Mathematical Puzzles in Parabola
David Angell and Thomas Britz

Mathematical Puzzles
Revised Edition
Peter Winkler

Mathematics of Tabletop Games
Aaron Montgomery

Puzzle and Proof
A Decade of Problems from the Utah Math Olympiad
Samuel Dittmer, Hiram Golze, Grant Molnar, and Caleb Stanford

For more information about this series please visit: https://www.routledge.com/AK-PetersCRC-Recreational-Mathematics-Series/book-series/RECMATH?pd=published, forthcoming&pg=2&pp=12&so=pub&view=list

Puzzle and Proof
A Decade of Problems from the Utah Math Olympiad

Samuel Dittmer, Hiram Golze,
Grant Molnar, and Caleb Stanford

CRC Press
Taylor & Francis Group
Boca Raton London New York

CRC Press is an imprint of the
Taylor & Francis Group, an **informa** business

AN A K PETERS BOOK

Designed cover image: Samuel Dittmer, Hiram Golze, Grant Molnar, and Caleb Stanford

First edition published 2025
by CRC Press
2385 NW Executive Center Drive, Suite 320, Boca Raton FL 33431

and by CRC Press
4 Park Square, Milton Park, Abingdon, Oxon, OX14 4RN

CRC Press is an imprint of Taylor & Francis Group, LLC

ISBN: 978-1-032-76263-0 (hbk)
ISBN: 978-1-032-75552-6 (pbk)
ISBN: 978-1-003-47776-1 (ebk)

DOI: 10.1201/9781003477761

Typeset in Latin Modern font
by KnowledgeWorks Global Ltd.

Publisher's note: This book has been prepared from camera-ready copy provided by the authors.

Contents

Preface   vii

Contributors   ix

CHAPTER   1 ▪ Introduction   1

1.1	HISTORY	1
1.2	THE PROBLEMS	1
1.3	GRADING AND RESULTS	2
1.4	STRUCTURE OF THIS BOOK	3
1.5	ACKNOWLEDGMENTS	3

CHAPTER   2 ▪ Problems   5

2.1	DISCRETE STRUCTURES	5
2.2	COUNTING AND PROBABILITY	8
2.3	GAMES	11
2.4	ALGEBRA	16
2.5	NUMBER THEORY	18
2.6	GEOMETRY	19

CHAPTER   3 ▪ Solutions   23

3.1	DISCRETE STRUCTURES	23
3.2	COUNTING AND PROBABILITY	33
3.3	GAMES	59
3.4	ALGEBRA	72
3.5	NUMBER THEORY	81
3.6	GEOMETRY	96

Bibliography   117

Index   119

Preface

This book is a collection of 70 mathematical problems and puzzles from the first 10 years of the Utah Math Olympiad (UMO), 2013–2022. These problems are distinguished in two respects. First, they aim to be *understandable* to an advanced high school audience, even if solving them can sometimes be quite difficult. Second, each problem asks, for not only an answer but also a *proof.*

You may be reading this book because you like proofs, or you may have never heard of them! A proof is a rigorous mathematical argument that demonstrates why a given answer is correct. Few high school students have exposure to proofs unless they go looking for trouble. In introductory geometry, you may have seen proofs as a fun diversion or as a frustrating exercise in demonstrating the obvious. Later, if you learned calculus, you moved on from proofs to proceed with the serious work of computing derivatives and integrals of functions that involve cosines and logarithms. But what is often not taught is that behind every concept and calculation—in geometry and calculus and beyond—there exists a proof that it works: a beautiful argument that, if understood, could convince even the most serious skeptic.

Why did we write this book? Simply put, to spread the beauty of mathematical proofs—to high schoolers or to anyone else. We hope this book will provide one more avenue for students to find trouble—should they go looking for it.

Samuel Dittmer, Hiram Golze, Grant Molnar, and *Caleb Stanford*

May 2024

Contributors

Peter Baratta
2012–2017

Samuel Dittmer
2012–present

Hiram Golze
2012–present

Wyatt Mackey
2014–2016

Grant Molnar
2015–present

Daniel South
2019–present

Caleb Stanford
2012–present

Josh Speckman
2019–2022

Annie Yun
2017–2020

Michael Zhao
2014–2015

CHAPTER 1

Introduction

1.1 HISTORY

We founded the Utah Math Olympiad (UMO) while studying at Brigham Young University in Fall 2012. The first contest was held on February 16, 2013, with 30 participants at two locations. The test had six problems to be solved in three hours.

We were motivated by our experience with and enthusiasm for mathematics competitions—including the American Mathematics Competitions series and the Utah State Math Contest. Yet, most entry-level contests such as these are limited to multiple-choice or short-answer problems. As a result, participants do not get exposure to proof-based mathematics. On the other hand, invitational proof contests like the USA Junior Math Olympiad (USAJMO) and USA Math Olympiad (USAMO) can be inaccessible to students who lack deep training in college-level mathematics; less than 500 students qualify for the USAJMO and USAMO combined every year.

We created the UMO to provide a bridge to mathematical writing for the rest—for high school students who are mathematically interested, but not yet familiar with writing proofs. Since then, the UMO has been held every year in the spring, from 2013 through 2024.

1.2 THE PROBLEMS

Of the 70 problems in this book, 60 are from the first 10 years of the UMO (6 problems per year). In addition, we also include 10 problems which we call "flyer problems." Every year, in order to spread the word about the contest, the committee has created a flyer featuring an original problem that is similar in style to the problems on the contest. This book compiles these problems and provides them with complete solutions for the first time.

The easier problems in each UMO contest (problems 1–3) are intended to be solvable using only the techniques in high school math prior to calculus. The harder problems (problems 4–6) are typically more challenging and may require clever applications of these elementary ideas. It is rare to be able

DOI: 10.1201/9781003477761-1

to solve all six problems—as of this writing, no student has ever received a perfect score. This is intentional! Some of the problems have even stumped the authors when they were first proposed. In general, some experience with high school math competitions is helpful—for example, exposure to concepts such as modular arithmetic, Euclidean geometry, and polynomials.

Our internal system uses six topics to categorize problems: *algebra, counting and probability, geometry, number theory, games,* and "other." The last two of these are nontraditional: the *games* problems require no background; they encourage you to play. The "other" problems don't fit into any of the previous topics, but for the purposes of this book, we have entitled this topic as *discrete structures.* These problems usually involve some sort of algorithm or else a manipulation of discrete objects.

All of the problems are selected by the problem-writing committee, which meets every spring to propose and discuss problems. Problems are proposed, then narrowed down to a shortlist and edited before selecting the final versions (which appear in this book). Out of 70 problems, 60 are credited to the present authors and 10 are credited to other contributors. We aim for problems that pique initial curiosity, are elegantly stated, and require creativity in exploring different avenues toward a solution.

1.3   GRADING AND RESULTS

Understanding how UMO problems are typically graded may offer useful context. Each problem is scored on a scale from 0 to 7, roughly as follows: (1) for a correct answer without proof, (2) for some progress, (3-4) for significant progress, (5) for a solution with significant holes, (6) for an almost perfect solution, and (7) for a flawless proof. A score of 5–7 can be considered "solving" the problem. Historically, the mean score on the UMO is 12.2: 4.0 points on problem 1, 3.1 points on problem 2, 2.4 points on problem 3, 0.6 points on problem 4, 0.5 points on problem 5, and 0.5 points on problem 6. A score of 10–14 out of 42 (two problems solved) is typically enough to receive a top-10 placement.

To ensure fairness, the UMO problem committee anonymizes the problem entries, and each problem is graded independently by at least two committee members. The committee members then reconcile their scores to give an official ruling. In addition, the committee identifies at most one "best solution" per problem, which is selected for clarity, rigor, and creativity.

The top 10 scorers each year are advertised on the Utah Math Olympiad website (utahmath.org). We also provide feedback to each student via email about any individual comments on their solutions provided by the graders. Cash prizes are awarded both to best solution winners and to the five highest scoring participants. The amounts of cash prizes are selected so as to follow the following somewhat unusual rule: the prize amount is always an *approximately linear* function of the score, while ensuring that each prize amount is a round multiple of 5 or 10.

1.4 STRUCTURE OF THIS BOOK

This book is organized into three chapters: Introduction, Problems, and Solutions. Within each chapter, the problems and solutions are grouped into six sections by the aforementioned topics: *discrete structures, counting and probability, games, algebra, number theory,* and *geometry.* Within each topic, the "flyer problems" appear first, followed by all other problems sorted by difficulty (from easiest to hardest), based on scores of participants who took the contest.

After trying a problem, you can find an explanation of its solution and proof in the corresponding solutions chapter. Some problems contain more than one solution, and there may even be other alternate solutions beyond the solutions we present. We also include the author of the problem, what fraction of test-takers were able to solve it (score 5–7), and its average (mean) score earned by that year's test-takers. These last two metrics provide rough measures of the problem's difficulty, but are omitted for the ungraded "flyer problems." The book concludes with a bibliography and an index for referencing, which you can use to look up specific proof techniques and topics.

Sometimes, the solutions will use terminology and techniques which you may not have seen before. If you want to learn more about the proof techniques that are used in this book—or are curious about any of the mathematical topics presented, including topics like combinatorics, number theory, and geometry—we recommend *The Art and Craft of Problem Solving* by Paul Zeitz [22] and the *Art of Problem Solving* online community and textbook series (aops.com).

If you get stuck on a problem, here are some tips. Make sure you know the meaning of all the terms used in the problem statement. Write down the definition of each one. Try a small example. For example, if the problem is about a variable n, try $n = 2$ or $n = 3$. Draw a picture of what the problem is describing. Work backward: if you had a solution, what would it look like? What steps of your proof could lead you to that point? Try thinking about the problem in a different way. Or try a *proof by contradiction:* what would happen if the statement were *not* true? If you can show that that is impossible, then the only possibility left is that the statement is true.

1.5 ACKNOWLEDGMENTS

This book would not be possible without the hard work of the Utah Math Olympiad problem-writing committee over the last 10, now 12, years. We would therefore like to thank Peter Baratta (2012–2017), Wyatt Mackey (2014–2016), Michael Zhao (2014–2015, in memoriam), Annie Yun (2017–2020), Daniel South (2019–present), Josh Speckman (2019–2022), and Zachary Klein (2024–present), as well as Benjamin Stanford (2015, problem 4). The collaborative process of proposing and working through problems, through many enjoyable calls on Google Hangouts (and later, Facebook Video Chat), has been an invaluable asset. Although not all problems have made the cut to appear on the contest and in this book, many others have had potential.

When the Utah Math Olympiad was in its infancy, our mentors at Brigham Young University, particularly Dr. David G. Wright and Dr. Tiancheng Ouyang, provided encouragement and inspiration. They even provided financial support for prize money during the first several years of the contest. Thanks to everyone who helped with proctoring; and to those who helped with test-solving, including: Matthew Babbitt, Steven Fan, Emil Geisler, Patrick McClintock, Kei Nishimura-Gasparian, Benjamin Pachev, Matthew Stanford, Peter Winkler, and others.

Last and foremost: thank you to the participants, who have moved on their own exciting careers, and some of whom have even joined the problem-writing committee. May your future endeavors be filled with the beauty of problem solving and with a relentless thirst for the truth.

CHAPTER 2

Problems

2.1 DISCRETE STRUCTURES

1. *(2017 Flyer)* When a square is subdivided into n rectangles, the resulting figure is called a *simple tiling* if there is no set of at least 2 (but not all n) of the rectangles which forms a larger rectangle. For example, here are simple tilings with 2 and 5 rectangles:

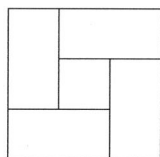

A *four-corners point* in a subdivision of a square into rectangles is a point where the corners of four rectangles meet.

Is there a simple tiling with a four-corners point?

2. *(2019 Flyer)* Lavender has six nickels, four of which are authentic and two of which are counterfeit. She knows that the counterfeit nickels weigh less than the authentic ones, but they weigh the same as each other. She also has a balancing scale.

A "move" using the scale consists of placing some number of nickels on each side of the scale; the scale will indicate either that the left side is heavier, that the right side is heavier, or that the two sides are equal.

Give a strategy to determine which two nickels are counterfeit which requires the fewest moves, and prove that no strategy with fewer moves is guaranteed to divide the counterfeit nickels from the authentic ones.

DOI: 10.1201/9781003477761-2

3. *(2017 P2)* Quinn places a queen on an empty 8×8 chessboard. She keeps the board secret, but she tells Alex the row that the queen is in, and she tells Adrian the column. Then she asks Alex and Adrian, alternately, whether or not they know how many moves are available to the queen.

- Alex says, "I don't know."
- Adrian says, "I didn't know before, but now I know."
- Alex says, "Now I know, too."

How many moves must be available to the queen? (Note: A queen can move to any square in the same row, column, or diagonal on the chessboard, except for its current square.)

4. *(2018 P1)*

   (a) Show that you can draw 7 circles and 11 dots on a page that satisfy all of the following properties:
   - No two circles intersect.
   - No dot lies on a circle.
   - Each circle encloses a different number of total dots (which could be 0).
   - Each dot is enclosed by the same number of total circles.

   (b) Prove that this is not possible with 5 circles and 5 dots.

5. *(2013 P1)* Consider the following diagram.

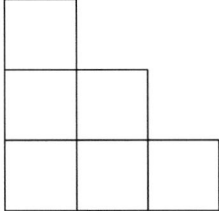

   (a) Show that you can retrace the diagram without lifting up your pencil using exactly nine (possibly overlapping) line segments.

   (b) Show that you cannot retrace the diagram in the same way using eight or fewer segments.

6. *(2020 P1)* An $n \times n$ grid of islands is connected by bridges, as in the following picture for $n = 3$:

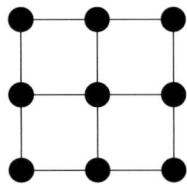

In the above picture, there are 9 islands and 12 bridges. A *path* consists of starting at any island, traveling along the bridges, and ending at any island. For example, a path could visit the three islands along the top of the grid from left to right, crossing two bridges along the way. A *perfect path* is a path such that:

- Every island is visited exactly twice;
- Every bridge is crossed at least once; and
- The path never makes a U-turn, i.e., it never travels along a bridge and then immediately back again.

(a) Find a perfect path when $n = 3$, or prove it is impossible.

(b) Find a perfect path when $n = 4$, or prove it is impossible.

7. *(2016 P3)* Can each positive integer $1, 2, 3, \ldots$ be colored either red or blue, such that for all positive integers a, b, c, d (not necessarily distinct), if $a + b + c = d$, then a, b, c, d are not all the same color?

8. *(2014 P4)* Joel is playing with ordered lists of integers in the following way. He starts out with an ordered list of nonnegative integers. Then, he counts the number of 0s, 1s, 2s, and so on in the list, writing the counts out as a new list. He stops counting when he has counted everything in the previous list. Then he takes the second list and applies the same process to get a third list. He repeats this process indefinitely.

For example, he could start out with the ordered list $(0, 0, 0, 2)$. He counts three 0s, zero 1s, and one 2, and then stops counting, so the second list is $(3, 0, 1)$. In the second list, he counts one 0, one 1, zero 2s, and one 3, so the third list is $(1, 1, 0, 1)$. Then he counts one 0 and three 1s, so the fourth list is $(1, 3)$. Here are the first few lists he writes down:

$$(0, 0, 0, 2) \longrightarrow (3, 0, 1) \longrightarrow (1, 1, 0, 1) \longrightarrow (1, 3) \longrightarrow \cdots$$

If instead he started with $(0, 0)$, he would write down:

$$(0, 0) \longrightarrow (2) \longrightarrow (0, 0, 1) \longrightarrow (2, 1) \longrightarrow \cdots$$

If Joel starts out with an arbitrary list of nonnegative integers and then continues this process, there are certain lists (m, n) of length two that he might end up writing an infinite number of times. Find all such pairs (m, n).

9. *(2022 P5)* 2022 lily pads are arranged in a circle. Each lily pad starts with height 1. A frog starts on one of the lily pads, and jumps around clockwise as follows: if the frog is on a lily pad of height k, the lily pad grows by 1 (becoming $k + 1$), and then the frog jumps k lily pads

clockwise (i.e., jumping over $(k-1)$). The frog continues doing this as long as it pleases.

After n jumps, let $D(n)$ be the difference between the tallest lily pad and the shortest lily pad. Find, with proof, the maximum possible value of $D(n)$, or prove that $D(n)$ is unbounded.

2.2   COUNTING AND PROBABILITY

1. *(2018 Flyer)* Consider 28 dots arranged in a triangular grid as shown, so that any two consecutive dots in a row together with the dot lying immediately above them forms an equilateral triangle.

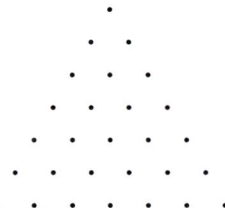

   How many distinct lines pass through exactly two dots of this grid?

2. *(2017 P1)* There are four cities and exactly one road between every pair of cities. How many ways can you paint the roads orange and blue such that it is possible to travel between any two cities on only orange roads, and it is also possible to travel between any two cities on only blue roads? Prove your answer.

   For example, the following coloring of the roads would *not* be allowed, because there is no way to travel between 1 and 2 on orange roads.

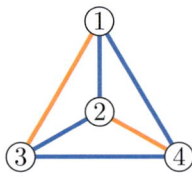

3. *(2019 P1)* How many ways can you divide a 4×4 square into a collection of one or more 1×1, 2×2, 3×3, and 4×4 squares? For example, three *different* ways are shown below.

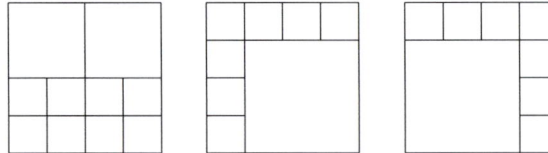

4. *(2020 P3)* In a 3×3 square grid, four of the nine squares are chosen at random and shaded. In the resulting figure, a *region* is a set of shaded squares that are vertically or horizontally (not diagonally) adjacent. For example, the following grid has two regions, one containing three squares and the other containing one square:

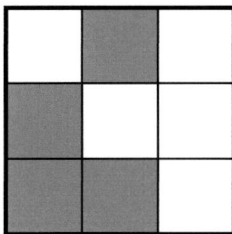

Find the expected value of the number of regions.

5. *(2016 P2)* Four fair six-sided dice are rolled. What is the probability that they can be divided into two pairs which sum to the same value? For example, a roll of $(1, 4, 6, 3)$ can be divided into $(1, 6)$ and $(4, 3)$, each of which sum to 7, but a roll of $(1, 1, 5, 2)$ cannot be divided into two pairs that sum to the same value.

6. *(2015 P5)* A 3×3 grid is filled with integers (positive or negative) such that the product of the integers in any row or column is equal to 20. For example, one possible grid is:

$$\begin{bmatrix} 1 & -5 & -4 \\ 10 & -2 & -1 \\ 2 & 2 & 5 \end{bmatrix}$$

In how many ways can this be done?

7. *(2014 P6)* Draw n rows of $2n$ equilateral triangles each, stacked on top of each other in a diamond shape, as shown in the following drawing when $n = 3$. Set point A as the southwest corner and point B as the northeast corner. A step consists of moving from one point to an adjacent point along a drawn line segment, in one of the four legal directions indicated. A path is a series of steps, starting at A and ending at B, such that no line segment is used twice. One path is drawn below. Prove that for every positive integer n, the number of distinct paths is a perfect square. (Note: A perfect square is a number of the form k^2, where k is an integer).

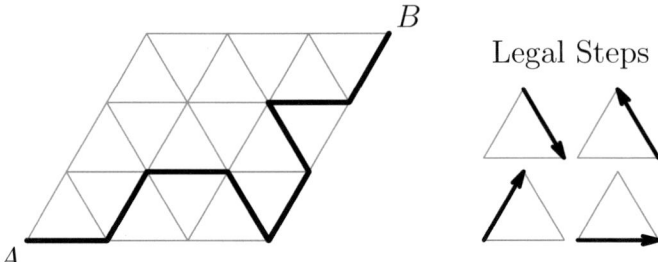

8. *(2018 P5)* Twelve labeled points are spaced equally on a circle, as shown below. How many ways are there to color 3 of the points red, 3 green, 3 blue, and 3 white such that for every two different colors, there exists a straight line that separates all the points of those two colors? (For example, there should be a line such that all the green points are on one side and all the white points are on the other.)

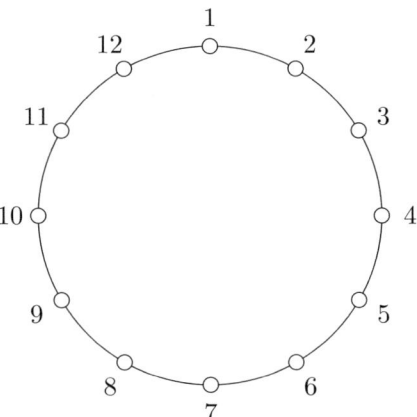

9. *(2021 P4)* Farmer Georgia has a positive integer number c of cows (which have four legs), and zero ostriches (which have two legs), on her farm on day 1. On each day thereafter, she adds a combination of cows and ostriches to her farm, so that on each day $n \geq 2$, the number of animals on the farm is equal to exactly half the number of legs that were on the farm on day $n - 1$. For example, there are $4c$ legs on day 1, so there must be exactly $2c$ animals on day 2. She may never remove animals from the farm.

Let $P_c(n)$ be the number of possible sequences of ordered pairs

$$(c_1, o_1), (c_2, o_2), \ldots, (c_n, o_n)$$

such that c_i and o_i are the number of cows and ostriches, respectively, on the farm on day i, where $(c_1, o_1) = (c, 0)$. For example, we have $P_1(2) = 2, P_1(3) = 5$, and $P_2(3) = 12$.

Find all positive integers c such that $P_c(2021)$ is a multiple of 3.

10. *(2022 P6)* An $m \times n$ grid of squares (with m rows and n columns) has some of its squares colored blue. The grid is called *fish-friendly* if a fish can swim from the left edge of the grid to the right edge of the grid only moving through blue squares. In other words, there is a sequence of blue squares, each horizontally or vertically adjacent to the previous square, starting in the first column and ending in the last column.

Prove that the number of fish-friendly 42×49 grids is at least 2^{2022}.

11. *(2019 P6)* Prove that for all $n \geq 200$, if $n \equiv 0$ or $n \equiv 1 \pmod 4$, then it is possible to put n balls of various colors in a bag such that when two balls are drawn out (without replacement), there is an equal probability of the two colors being the same and the two colors being different.

For example, this is possible when $n = 13$: the bag can contain 9 red balls, 3 green balls, and 1 blue ball.

12. *(2017 P6)* In a deck of n cards, there is one card of each number from 1 to n. Let a_n be the number of orderings of the deck such that the first card is less than the second card, the second card is greater than the third card, the third card is less than the fourth card, and so on. For example, $a_1 = 1$, $a_2 = 1$, $a_3 = 2$, and $a_4 = 5$.

Determine, for all n, the remainder when a_n is divided by 4.

13. *(2013 P6)* How many ways can one tile the border of a triangular grid of hexagons of length n completely using only 1×1 and 1×2 hexagon tiles? Express your answer in terms of a well-known sequence, and prove that your answer holds true for all positive integers $n \geq 3$. Examples of such grids for $n = 3$, $n = 4$, $n = 5$, and $n = 6$ are shown below.

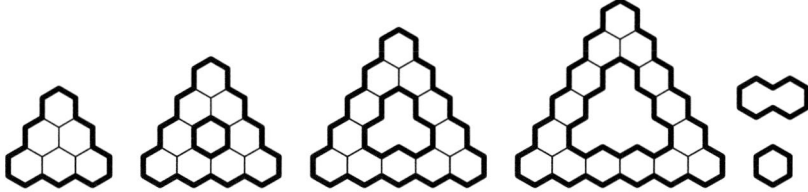

2.3 GAMES

1. *(2016 Flyer)* Malcolm and Ozymandias are playing a modified version of Tic-Tac-Toe. Each turn, they place one of the following numbers:

$$2, 3, 4, 5, 6, 8, 9, 10, 15$$

into a Tic-Tac-Toe board. No number can be used more than once. Malcolm wins if three numbers in the same row, column, or diagonal are pairwise relatively prime. (*Pairwise* relatively prime means that every pair of two out of the three numbers is relatively prime.) Ozymandias wins if this does not happen. Malcolm goes first. Which player has a winning strategy?

2. *(2021 Flyer)* If m and n are integers, we say that m covers n if m is a multiple of n. In particular, m covers itself. Bonnie and Clyde play the following game.

   Players alternately take turns, with Bonnie taking the first turn. On her first turn, Bonnie names an integer from 1 to 10 inclusive. Thereafter, each player names another integer from 1 to 10 that is not covered by any previously named integer. When a player is unable to name such an integer, he or she loses the game.

   If both Bonnie and Clyde play optimally, who wins the game? Describe the winning strategy for that player.

3. *(2014 P1)* Todd and Allison are playing a game on the grid shown below. At the beginning, an orange stone is placed in the center intersection on the grid. They take turns, with Todd going first. In each of Todd's turns, he must move the orange stone from its current position to a horizontally or vertically adjacent intersection that is not occupied by a blue stone, and then he places a blue stone in the orange stone's previous spot. In each of Allison's turns, she places a blue stone on exactly one unoccupied intersection. Todd loses the game when he is forced to move into one of the corner intersections, labeled by A, B, C, and D in the diagram below. Allison loses if Todd can't move.

   Allison tries to force Todd to lose in as few as turns as possible, and Todd tries to survive as long as possible. If both of them play as best they can, how many blue stones will be on the board at the end of the game? (You may assume that Todd always loses.)

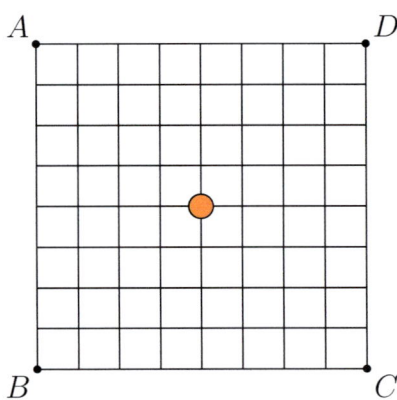

4. *(2016 P1)* Ada and Otto are engaged in a battle of wits. In front of them is a figure with six dots, and nine sticks are placed between pairs of dots as shown below. The dots are labeled A, B, C, D, E, and F. Ada begins the game by placing a pebble on the dot of her choice. Then, starting with Ada and alternating turns, each player picks a stick adjacent to the pebble, moves the pebble to the dot at the other end of the stick, and then removes the stick from the figure. The game ends when there are no sticks adjacent to the pebble. The player who moves last wins. A sample game is described below.

If both players play optimally, who will win?

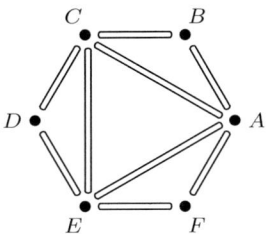

Sample Game

1. Ada places the pebble at B.

2. Ada removes the stick BC, placing the pebble at C.

3. Otto removes the stick CD, placing the pebble at D.

4. Ada removes the stick DE, placing the pebble at E.

5. Otto removes the stick EA, placing the pebble at A.

6. Ada removes the stick AB and wins.

5. *(2018 P3)* Kim and Li play the following game. Kim places a 2×1 rectangle vertically somewhere on the 4×4 grid below. Then Li places a 1×2 rectangle on the grid horizontally so that it does not overlap with Kim's rectangle. These plays must be in-line with the grid, so no rectangle can partially cover more than two squares on the grid. Play repeats in this fashion until one of the players is unable to place any more rectangles. The last player to move wins.

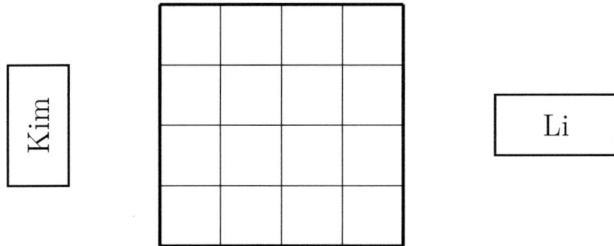

If both Alpha and Beta play optimally, determine, with proof, who wins.

6. *(2019 P4)* A game is played between a "leader," a ball, and 10 "players" (not including the leader). One of the players is secretly the traitor. The leader begins with the ball, and passes it to one of the players. When a player receives the ball, they take a turn, which consists of passing the ball to another player (not themselves and not the leader). Players continue taking turns according to who has the ball.

Before the game begins, the leader gives *instructions* to each player. (The leader is allowed to give different instructions to different players; see the example below.) The instructions consist of an infinite list of player names. When a player receives the ball for the 1st time, they pass the ball to the 1st player on their list; when they receive the ball for the 2nd time, they pass the ball to the 2nd player on their list, and so on. All players follow the instructions exactly, except the traitor, who may ignore the instructions. (Assume that a player's instructions do not include their own name.)

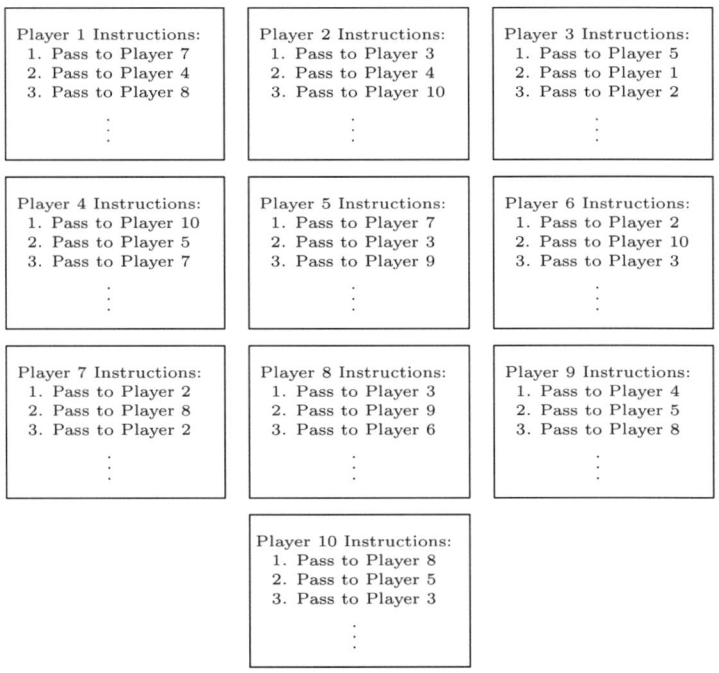

If the ball is passed to every player at least once (including the traitor), the game ends and the leader wins. On the other hand, if the ball is passed twice to the traitor, the game ends and the traitor wins. If the game goes on forever, the traitor also wins.

Prove that the leader may give instructions to the players so that the leader wins no matter which player is the traitor and how the traitor plays; or prove that this is impossible.

7. *(2013 P2)* Alice and Carl play the following game using a square sheet of paper. On each turn, the player makes a straight cut through the sheet (not necessarily parallel to the sides of the page), creating two new sheets. The sheet with smaller area is discarded (either one if the two are equal), and the player gives the larger sheet to the other player. The first player to receive a sheet of area less than 1 square centimeter from the opposing player loses.

If Alice goes first, A the initial area of the paper, and $A \geq 1$, describe (with proof) all values of A for which she has a winning strategy.

8. *(2017 P5)* The Great Pumpkin challenges you to the following game. In front of you are 8 empty buckets. Each bucket is able to hold 2 liters of water. You and the Great Pumpkin take turns, with you going first.

On your turn, you take 1 liter of water and distribute it among the buckets in any amounts you like. On the Great Pumpkin's turn, it empties two of the buckets of its choice. The Great Pumpkin is defeated if you cause any of the buckets to overflow.

Given sufficiently many turns, can you defeat the Great Pumpkin no matter how it plays? Prove it, or prove that it is impossible.

9. *(2021 P5)* Gog and Magog are playing a game with stones. Each player starts out with no stones, and they alternate taking turns. Gog goes first. On each turn, a player can either gain one new stone, or give at least one and no more than half of their stones to the other player. If a player has 20 or more stones, they lose.

Determine, with proof, whether Gog has a winning strategy, Magog has a winning strategy, or neither player has a winning strategy (the game goes on indefinitely).

10. *(2022 P4)* Alpha and Beta are playing a game on a 10×100 grid of squares. At each turn, they can fold the grid along any of the interior horizontal or vertical gridlines, which creates a smaller (folded) grid of squares (on the first move, they can choose one of 9 horizontal or 99 vertical gridlines). The person who makes the last fold wins.

If both players play optimally and Alpha starts, determine, with proof, who wins.

11. *(2020 P6)* The positive integers between 1 and 10 are holding an election. They are sitting around a circular table—1, then 2, then 3, and so on in clockwise order. Starting with 1 and going clockwise, each integer votes

for a president (between 1 and 10). After all 10 integers have voted, the player with the most votes wins the election. The higher integer wins in case of tie.

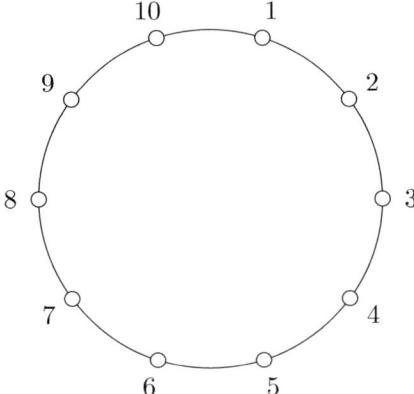

Every integer prefers itself to win; but if it can't win, it prefers the other integers in clockwise order from itself. For example, 8 prefers itself, then 9, then 10, then 1, then 2, and so on. Every integer behaves perfectly rationally and knows that every other integer will behave perfectly rationally as well.

Who wins the election? Prove your answer.

12. *(2013 P5)* Cooper and Malone take turns replacing a, b, and c in the equation below with real numbers.

$$P(x) = x^3 + ax^2 + bx + c$$

Once a coefficient has been replaced, no one can choose to change that coefficient on their turn. The game ends when all three coefficients have been chosen. Malone wins if $P(x)$ has a nonreal root, and Cooper wins otherwise.

If Malone goes first, find the person who has a winning strategy, and describe it with proof.

2.4   ALGEBRA

1. *(2014 Flyer)* A $m \times n$ matrix of nonnegative real numbers is called "balanced" if the average of the values in any row or column is equal to 1. Find the maximum possible value of the **minimum nonzero element** in a balanced 4×5 matrix.

For example, the following is an example of a balanced 3×5 matrix with minimum nonzero element 1.25.

$$\begin{bmatrix} 1.5 & 1.75 & 1.75 & 0 & 0 \\ 0 & 1.25 & 1.25 & 1.25 & 1.25 \\ 1.5 & 0 & 0 & 1.75 & 1.75 \end{bmatrix}$$

2. *(2020 Flyer)* Define the sequence $A_1, A_2, \ldots$ by setting $A_1 = 1$, $A_2 = 2$, and

$$A_n = \left(\frac{n^2}{2} - \frac{2}{(n-1)^2} \right) A_{n-1} + A_{n-2}$$

for all $n > 2$. Find and prove an explicit formula for A_n in terms of one or more well-known sequences.

3. *(2022 Flyer)* Let a, b, and c be real numbers such that $a2^n + bn + c$ is an integer for all nonnegative integers n. Must a, b, and c be integers?

4. *(2018 P2)* Let $P(x)$ be a cubic polynomial $(x - a)(x - b)(x - c)$, where a, b, and c are positive real numbers. Let $Q(x)$ be the polynomial with $Q(x) = (x - ab)(x - bc)(x - ca)$. If $P(x) = Q(x)$ for all x, then find the minimum possible value of $a + b + c$.

5. *(2021 P1)* Find all ordered triples of integers (x, b, c), such that b is prime, c is odd and positive, and $x^2 + bx + c = 0$.

6. *(2019 P2)* Find all polynomials $p(x)$ with real coefficients such that $p(x)^2 = p(p(x))$.

7. *(2022 P3)* Find all sequences $a_1, a_2, a_3, \ldots$ of real numbers such that for all positive integers $m, n \geq 1$, we have

$$a_{m+n} = a_m + a_n - mn \text{ and}$$
$$a_{mn} = m^2 a_n + n^2 a_m + 2a_m a_n.$$

8. *(2021 P3)* To each point P in the plane, a real number $f(P)$ is assigned. Is it possible that for every equilateral triangle PQR in the plane, $f(P) + f(Q) + f(R)$ is equal to the perimeter of $\triangle PQR$?

9. *(2015 P4)* Anastasia and Balthazar need to go to the grocery store, which is 100 km away. Anastasia walks at 5 km/hr and Balthazar walks at 4 km/hr. However, they also own a single bike, and each of them bikes at 10 km/hr. They are allowed to go forward or backward, and the bike will not get stolen if they drop it off along the way for the other person to pick up. What is the shortest amount of time necessary for both of them to get to the grocery store?

10. *(2020 P2)* Let a and b be real numbers with the property that $a^n - b^n$ is rational for every positive integer $n \geq 2$. Show that either a and b are both rational, or that $a = b$.

11. *(2016 P5)* Let $a_0, a_1, a_2, \ldots$ be a sequence of integers (positive, negative, or zero) such that for all nonnegative integers n and k,

$$a_{n+k}^2 - (2k+1)a_n a_{n+k} + (k^2+k)a_n^2 = k^2 - k.$$

Find all possible sequences (a_n).

12. *(2014 P5)* Find all positive real numbers x, y, and z that satisfy both of the following equations:

$$xyz = 1$$
$$x^2 + y^2 + z^2 = 4x\sqrt{yz} - 2yz.$$

2.5 NUMBER THEORY

1. *(2013 Flyer)* Let $f(n)$ be the number of positive integers k such that 2^k is less than or equal to 3^n. Prove that $f(n+1) - f(n) = 1$ for infinitely many positive integers n.

2. *(2015 P3)* Find, with proof, all positive integers n with $2 \leq n \leq 20$ such that the greatest common divisor of the coefficients of $(x+y)^n - x^n - y^n$ is equal to exactly 3.

3. *(2022 P2)* Let x and y be relatively prime integers. Show that $x^2 + xy + y^2$ and $x^2 + 3xy + y^2$ are relatively prime.

4. *(2015 P1)* Three trolls have divided n pancakes among themselves such that:

   • Each troll has a positive integer number of pancakes.
   • The greatest common divisor of the number of pancakes held by any two trolls is bigger than 1.
   • The three greatest common divisors obtained in this way are all distinct.

   What is the smallest possible value of n?

5. *(2019 P3)* Find all nonnegative integers a, b, and c such that $2^a + 2^b = c!$.

6. *(2014 P2)*

   (a) Find all positive integers x and y that satisfy $x^2 + y^2 = 2014$, or prove that there are no solutions.

   (b) Find all positive integers x and y that satisfy $x^2 + y^2 = 3222014$, or prove that there are no solutions.

7. *(2020 P5)* We say a triangle with integer side lengths a, b, and c is *primitive* if a, b, and c share no common factor greater than 1, and *special* if it has an angle with measure $120°$. For example, the following triangle with side lengths 3, 5, and 7 is both primitive and special:

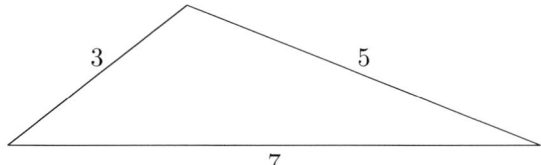

Prove that there are infinitely many primitive special triangles.

8. *(2013 P3)* Find all x with $1 \leq x \leq 999$ such that the last three digits of x^2 are all equal to the same nonzero digit.

9. *(2017 P4)* The polynomial $P(x)$ has integer coefficients, and satisfies

$$n\, P(n) \equiv 1 \pmod{16}$$

for every odd positive integer n. That is, $nP(n)$ is 1 greater than a multiple of 16. Find, with proof, the minimum possible degree of $P(x)$.

10. *(2021 P6)* Prove that for all positive integers n, the number of divisors of $n!$ is a divisor of $(2n)!$.

11. *(2016 P6)* Find all positive integer pairs (u, m) such that $u + m^2$ is divisible by $um - 1$.

12. *(2018 P6)* If p and q are distinct prime numbers, then determine all possible values of

$$\gcd\left(p - 1, \frac{q^p - 1}{q - 1}\right).$$

(For positive integers x and y, $\gcd(x, y)$ denotes the greatest common divisor of x and y.)

2.6   GEOMETRY

1. *(2015 Flyer)* A fractal figure is formed as follows. First, draw a 45-45-90 right triangle with a hypotenuse of length 1. Then draw two new 45-45-90 triangles whose hypotenuses are the first right triangle's legs, and whose legs lie outside the first triangle. Then, for each of the two new triangles, do the same thing, so that there are two new right triangles for each right triangle in the previous iteration. Repeat this process to infinity. The diagram below shows the figure after three iterations of the process:

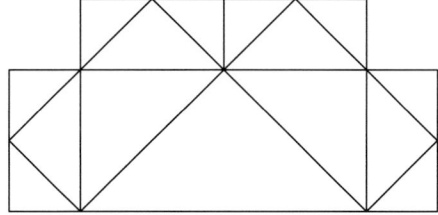

    (a) Prove that the resulting figure has finite area. (If triangles overlap, their common area is only counted once.)

    (b) Find this area.

2. *(2022 P1)* Let $n \geq 2$ be an integer. Thibaud the Tiger lays n 2×2 overlapping squares out on a table, such that that the centers of the squares are equally spaced along the line $y = x$ from $(0, 0)$ to $(1, 1)$ (including the two endpoints). For example, for $n = 4$ the resulting figure is shown below, and it covers a total area of $\frac{23}{3}$.

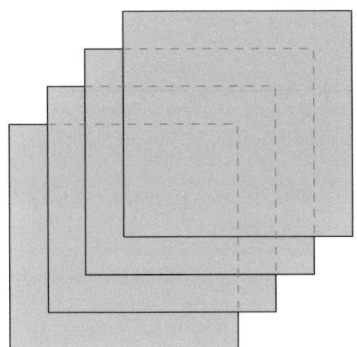

Find, with proof, the minimum n such that the figure covers an area of at least $\sqrt{63}$.

3. *(2015 P2)* In triangle ABC, $AC = 13$, $AB = 20$, and the length of the altitude from A to $\overleftrightarrow{BC}$ is 12. If M is the midpoint of $\overline{BC}$, find all possible length(s) of AM, and demonstrate that these length(s) are achievable.

4. *(2017 P3)* Three distinct points A, B, and C lie on a circle with center O, such that $\angle AOB$ and $\angle BOC$ each measure 60 degrees. Point P lies on chord (line segment) $\overline{AC}$ and the triangle BPC is drawn. If P is chosen so that it maximizes the length of the altitude from C to $\overline{BP}$, then determine the ratio $AP : PC$.

5. *(2021 P2)* Three circles, C_1, C_2, and C_3, are drawn in the plane such that each pair is externally tangent. Circle D is drawn externally tangent to all three, and circle E internally tangent to all three. If D and E have the same center, prove or disprove that C_1, C_2, and C_3 must have the same radius.

6. *(2018 P4)* Square $ABCD$ lies inside triangle XYZ such that B is on $\overline{XY}$, C is on $\overline{YZ}$, and D is on $\overline{ZX}$. Segment $\overline{XA}$ is also drawn. The result is four triangular regions ($\triangle XAB$, $\triangle YBC$, $\triangle ZCD$, and $\triangle XDA$), each adjacent to one of the four sides of the square. Prove that if these four triangular regions are each reflected across the adjacent side of the square, the resulting four reflected regions completely cover the square.

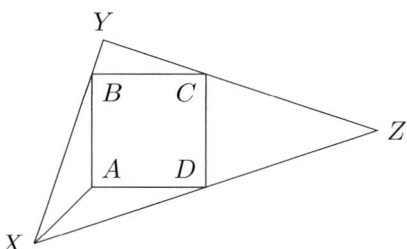

7. *(2016 P4)* Equiangular hexagon $ABCDEF$ has $AB = CD = EF$ and $AB > BC$. Segments $\overline{AD}$ and $\overline{CF}$ intersect at point X and segments $\overline{BE}$ and $\overline{CF}$ intersect at point Y. If quadrilateral $ABYX$ can have a circle inscribed inside of it (meaning there exists a circle that is tangent to all four sides of the quadrilateral), then find $\frac{AB}{FA}$.

8. *(2014 P3)* Completely describe the set of all right triangles with positive integer-valued legs such that when four copies of the triangle are arranged in square formation shown below, the incenters of the four triangles lie on the extensions of the sides of the smaller square.

   Note: the *incenter* of a triangle is the center of the circle inscribed in that triangle.

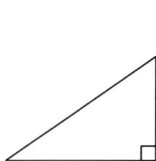 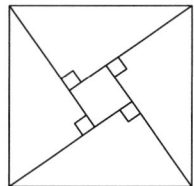

9. *(2020 P4)* Consider a circle C and a line L which does not intersect C. Let A be the point on C nearest to L and let B be the point on C furthest from L; let line AB intersect L at point X. Let C' be the circle centered at B passing through X, and let K be an intersection of C' with the perpendicular bisector of line segment $\overline{AX}$. Finally, let the circle with center X passing through K intersect line segment $\overline{AB}$ at a point H. Prove that for every point P lying on L, the circle through H centered at P is perpendicular to C.

   Note: Two intersecting circles C_1 and C_2 are called *perpendicular* if they intersect at two right angles. In other words, for each intersection point I of C_1 and C_2, the tangent lines to the circles at I are perpendicular.

10. *(2013 P4)* Given a line L_1 and distinct points I and X on L_1, draw lines L_2 and L_3 through point I, with angles α and β as marked in the figure. Also, draw line segment XY at an angle of γ from line L_1 such that it intersects line L_2 at Y. Establish necessary and sufficient conditions on

α, β, and γ such that a triangle can be drawn with one of its sides as XY with lines L_1, L_2, and L_3 as the angle bisectors of that triangle.

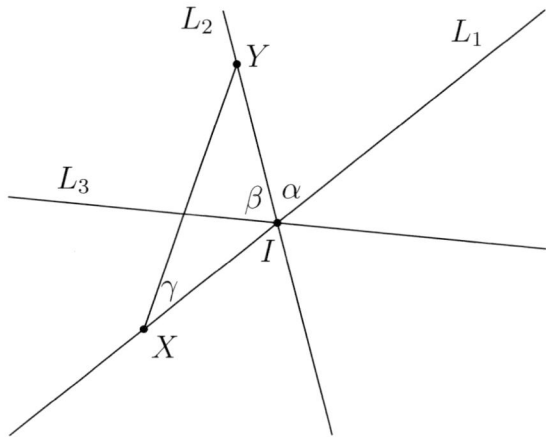

11. *(2015 P6)* A triangular pyramid with apex O and base ABC has the property that the perimeter of triangle ABC is 84. Additionally, one can place a cylinder of radius 4 and height 10 completely inside the pyramid such that one of its bases is in the same plane as triangle ABC. What is the minimum possible height from triangle ABC to apex O? Show that this height is achievable.

12. *(2019 P5)* Let I be the incenter of triangle ABC and D the point on $\overline{AB}$ tangent to the incircle. The line ID meets the circumcircle of ABI at I and E. If $IE = AC + BC$, what is angle C?

Note: *incircle* of a triangle is the circle inscribed inside it, and the *circumcircle* of a triangle is the circle through its three vertices. The *incenter* is the center of the incircle.

CHAPTER 3

Solutions

3.1 DISCRETE STRUCTURES

1. *(2017 Flyer, Author: Caleb Stanford)*

   Yes. The following is a simple tiling with a four-corners point:

   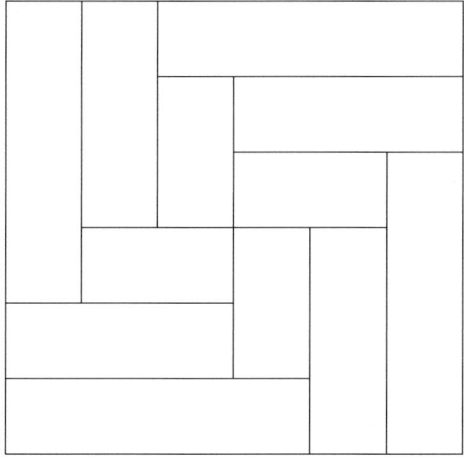

   Note: This problem was inspired by a puzzle posted on Stack Exchange by Owen Wilson [6].

2. *(2019 Flyer, Author: Grant Molnar)*

   Lavender can find the two counterfeit nickels in 3 moves, and this is optimal.

   First, we demonstrate a strategy that uses 3 moves. Begin by placing three nickels on each side of the scale.

   - If one side weighs less, we know that both counterfeit nickels must lie on that side of the scale. Then we weigh two of these three coins

DOI: 10.1201/9781003477761-3

against each other. If they are equal they are both counterfeit, and otherwise the lesser and the remaining nickel are counterfeit. In this case the strategy takes only 2 moves.

- Otherwise, we have divided the nickels into two groups of three, and one counterfeit nickel must be in each group. In each group, we can identify the counterfeit nickel in one move by weighing two of the three coins against each other. If one nickel weighs less it is counterfeit, otherwise the remaining nickel is counterfeit. So we can identify both counterfeit nickels using 2 additional moves, for a total of 3 moves.

Now we claim that this is optimal: Lavender can't identify the counterfeit nickels in only 2 moves. To prove this, suppose the six nickels are arranged in any fixed order, and consider any strategy that uses two moves. Each move has 3 outcomes (greater than, less than, or equal), so there are at most $3^2 = 9$ total ways that the strategy can play out. But the number of ways we can arrange two counterfeit nickels out of six is $\binom{6}{2} = 15$, which is greater than 9. By the pigeonhole principle, there must exist two arrangements for which the strategy plays out the same way, and so these two arrangements are indistinguishable at the end of the process. Therefore, Lavender cannot know which two coins are counterfeit using only 2 moves.

Note: This problem is a variation on the well-known class of "counterfeit coin problems" [14, 15].

3. *(2017 P2, Solved By: 94%, Average Score: 6.4, Author: Caleb Stanford)*

The queen has 23 moves available.

First, we can label each square on the chessboard by how many moves the queen has if it is on that square, as shown below.

21	21	21	21	21	21	21	21
21	23	23	23	23	23	23	21
21	23	25	25	25	25	23	21
21	23	25	27	27	25	23	21
21	23	25	27	27	25	23	21
21	23	25	25	25	25	23	21
21	23	23	23	23	23	23	21
21	21	21	21	21	21	21	21

Suppose that Quinn told Alex that the queen is in row 1 or row 8. Then Alex would immediately know how many moves the queen has. Therefore, the only way to satisfy Alex's first statement is if the queen is in one of rows 2–7.

Using the first part of Adrian's statement, we find by the same logic that the queen must be in one of columns 2–7. So before Adrian heard Alex, Adrian knew which column the queen was in (one of columns 2–7), but after learning that the queen was in one of rows 2–7, Adrian knew how many moves the queen had. In more detail, we find the following:

- If Adrian knows that the queen is in column 1 or 8, then (as we just noted) Adrian would immediately know that the queen has 21 moves before Alex's statement, a contradiction.

- If Adrian knows that the queen is in column 2 or 7, then the queen could have either 21 or 23 moves available. However, when Adrian learns that the queen cannot be in rows 1 or 8, then Adrian finds that the queen must have 23 moves. This seems to fit with the given statements.

- If Adrian knows that the queen is in column 3 or 6, then the queen could have 21, 23, or 25 moves available. When Adrian learns that the queen cannot be in rows 1 or 8, Adrian is only able to conclude that the queen has either 23 or 25 moves, hence the second statement would not make sense in this case.

- If Adrian knows that the queen is in column 4 or 5, then the queen could have 21, 23, 25, or 27 moves available. When Adrian learns that the queen cannot be in rows 1 or 8, Adrian is only able to conclude that the queen has 23, 25, or 27 moves, hence the second statement would not make sense in this case.

Therefore, we see that the only way that the first two statements make sense are if the queen is in column 2 or column 7, in one of rows 2–7. Therefore, the queen has 23 moves available.

Note that the third statement is unnecessary, as Alex is simply confirming the what we know as observers.

4. *(2018 P1, Solved By: 60%, Average Score: 4.3, Author: Caleb Stanford)*

   (a) The following diagram demonstrates that the desired configuration can be achieved.

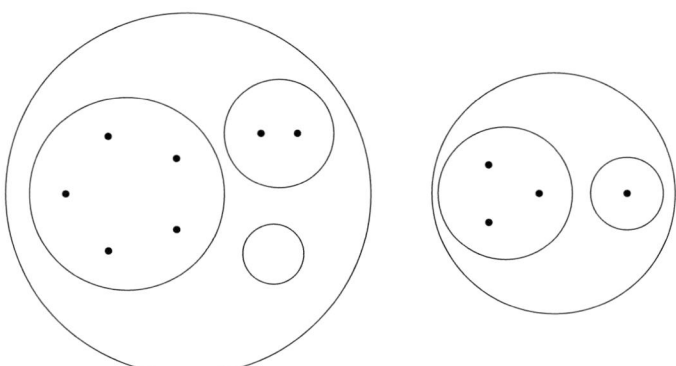

In the diagram, every dot is enclosed by exactly 2 circles. The circles on the left enclose 0, 2, 5, and 7 dots, and the circles on the right enclose 1, 3, and 4 dots; all of these values are unique.

One way to come up with this diagram is by a counting argument. We will count the number of ordered pairs (C, D) where C is a circle and D is a dot contained inside of C. Suppose that each dot is enclosed in k circles. Then the number of such pairs (C, D) is k for each circle, or $11k$.

Alternatively, suppose that the number of dots contained in circle i is x_i, so then $x_1, x_2, \ldots, x_7$ must all be distinct. Then the number of pairs (C, D) is

$$11k = x_1 + x_2 + x_3 + x_4 + x_5 + x_6 + x_7.$$

Therefore, we want the sum of seven distinct numbers to be a multiple of 11. The smallest possible sum of seven distinct nonnegative integers is $0 + 1 + 2 + \cdots + 6 = 21$, so we try $k = 2$, with $\{x_1, x_2, \ldots, x_7\} = \{0, 1, 2, 3, 4, 5, 7\}$. In other words, we look for a diagram where each dot is contained in two circles, and the circles contain $0, 1, 2, 3, 4, 5,$ and 7 dots. From here, we can come up with the diagram given in the beginning.

(b) For the sake of contradiction, assume that it is possible to draw five dots and five circles with these properties. By the same logic as in part (a), if each dot is enclosed in k circles, and if circle i contains x_i dots, then

$$5k = x_1 + x_2 + x_3 + x_4 + x_5.$$

We also know that the x_i are distinct nonnegative integers with $x_i \leq 5$ (because a circle cannot contain more than 5 dots in it).

Therefore,

$$10 = 0 + 1 + 2 + 3 + 4$$
$$\leq x_1 + x_2 + x_3 + x_4 + x_5$$
$$\leq 1 + 2 + 3 + 4 + 5$$
$$= 15,$$

which implies that $10 \leq 5k \leq 15$, so $k = 2$ or $k = 3$. We need to show both cases are impossible.

- If $k = 2$, then $x_1 + x_2 + x_3 + x_4 + x_5 = 10$, and the only way that we can choose the x_i to be distinct is if $\{x_1, x_2, x_3, x_4, x_5\} = \{0, 1, 2, 3, 4\}$.
- If $k = 3$, then $x_1 + x_2 + x_3 + x_4 + x_5 = 15$, and the only way that we can choose the x_i to be distinct with $0 \leq x_i \leq 5$ is if $\{x_1, x_2, x_3, x_4, x_5\} = \{1, 2, 3, 4, 5\}$.

Either way, there exists circles with 4 dots, 3 dots, and 2 dots: call them C_4, C_3, and C_2. Note that C_3 must lie inside C_4, and C_2 must in turn lie inside C_3 (since there are only 2 dots outside of C_3, and they are separated by C_4). See the following diagram:

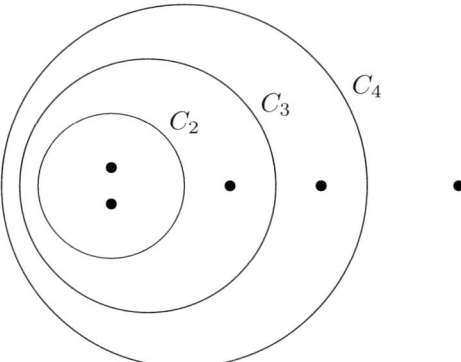

Now the point inside C_4 and outside C_3 has two fewer circles around it than the points inside C_2, so to make up, it must have at least two circles around it and outside of C_3. But then these two circles each contain only one dot, which is a contradiction. Therefore, it is impossible to draw a valid diagram with 5 circles and 5 dots.

5. (*2013 P1, Solved By: 37%, Average Score: 4.1, Author: Hiram Golze*)

(a) The path below that starts at B and ends at E and travels along the segments in the direction of the arrows retraces the diagram in exactly nine segments without lifting the pencil from the page.

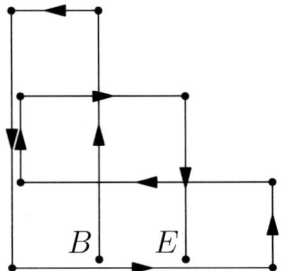

(b) Suppose we dissect the diagram into two pieces as shown below:

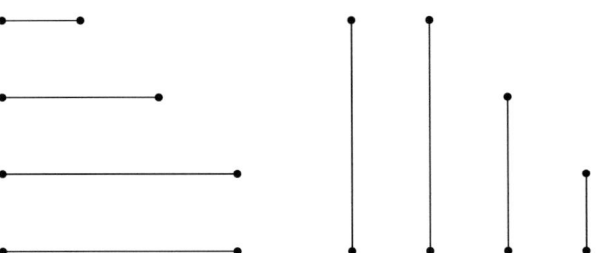

Note that we have eight distinct segments, so if we had a retracing that used exactly eight segments, then these would have to be our segments (connected in some order). The dots shown above are the only locations where we can change direction. If we are counting the number of dots in a valid path, we will count one for the starting point, seven for the changes in direction, and one for the ending point, for a total of nine dots. If we overlay the two diagrams above, however, we count ten distinct dots that we must visit:

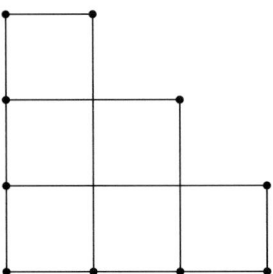

This is a contradiction, so we cannot retrace the diagram using eight segments.

6. *(2020 P1, Solved By: 47%, Average Score: 3.8, Author: Caleb Stanford)*

(a) A perfect path is impossible. Color the islands red and blue in a checkerboard pattern, as in the following picture:

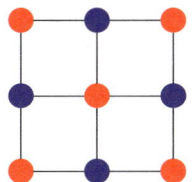

Any path must alternate between red and blue islands. Therefore, the number of visits to red islands and the number of visits to blue islands must either be equal, or off by 1. However, since a perfect path must visit every island twice, it must have 10 visits to red islands and 8 visits to blue islands, which is impossible.

(b) The following is one possible perfect path.

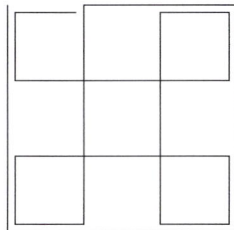

Alternatively, we can represent it using xy-coordinates:

$$(0,3) \to (0,0) \to (3,0) \to (3,3) \to (1,3) \to (1,0)$$
$$\to (0,0) \to (0,1) \to (3,1) \to (3,0) \to (2,0) \to (2,3)$$
$$\to (3,3) \to (3,2) \to (0,2) \to (0,3) \to (1,3).$$

Note: In general, a perfect path is possible whenever n is even, but not when n is odd.

7. *(2016 P3, Solved By: 33%, Average Score: 2.5, Author: Caleb Stanford)*

We claim that no such coloring exists. For the sake of contradiction, assume that such a coloring exists. Without loss of generality, we may assume that 1 is colored red. In the string of steps below, if a number n is known to be colored red, we denote it as n^R, and if it is known to be colored blue, we denote it as n^B. Each of the implications below follow from the fact that if all but one of the distinct numbers in $a + b + c = d$ are the same color, then the final number must be a different color.

$$1^R + 1^R + 1^R = 3 \quad \implies \quad 3 \text{ is blue.}$$
$$3^B + 3^B + 3^B = 9 \quad \implies \quad 9 \text{ is red.}$$
$$1^R + 4 + 4 = 9^R \quad \implies \quad 4 \text{ is blue.}$$

Now we have

$$1^R + 1^R + 9^R = 11 \quad \implies \quad 11 \text{ is blue}$$

but also,

$$3^B + 4^B + 4^B = 11 \quad \implies \quad 11 \text{ is red.}$$

This is a contradiction because 11 can only have one color. Therefore, no such coloring exists.

8. *(2014 P4, Solved By: 15%, Average Score: 1.8, Author: Hiram Golze)*

There are only 3 such pairs: $(1, 2), (2, 2)$, and $(0, 3)$.

If Joel starts out with a list whose length is not 2, and eventually encounters a list (m, n) of length 2, he might as well have started out with the list (m, n) (because we only care about what lists appear an infinite number of times). On the other hand if he never encounters a list (m, n) of length 2, we can safely ignore this case as no lists of length 2 will appear infinitely often. Therefore, we may assume that Joel starts out with a list of length 2.

We claim that for any $m, n \geq 0$, $(m, n) \to \cdots$ eventually arrives at one of the following two repeating patterns:

$$(1, 2) \to (0, 1, 1) \to (1, 2) \to \cdots \tag{$*$}$$
$$(2, 2) \to (0, 0, 2) \to (2, 0, 1) \to (1, 1, 1)$$
$$\to (0, 3) \to (1, 0, 0, 1) \to (2, 2) \to \cdots \tag{$**$}$$

Clearly, this is true for the lists $(m, n) = (1, 2), (2, 2)$, and $(0, 3)$. Next, consider the case where $(m, n) = (k, 2)$ or $(2, k)$, for $k > 2$:

$$\begin{matrix} (k, 2) \to \\ (2, k) \to \end{matrix} \ (0, 0, 1, \underbrace{0, 0, \ldots, 0}_{k-3 \text{ zeros}}, 1) \to (k - 1, 2)$$

Then $(k - 1, 2)$ will eventually become $(k - 2, 2)$, and so on, until we arrive at $(2, 2)$, which will repeat pattern $(**)$.

Next, consider all other cases where $m = 2$ or $n = 2$ not yet covered:

$$\begin{matrix} (2, 1) \to \\ \end{matrix} (0, 1, 1) \to (1, 2)$$
$$\begin{matrix} (2, 0) \to \\ (0, 2) \to \end{matrix} (1, 0, 1) \to (1, 2)$$

and $(1, 2)$ repeats pattern $(*)$.

We now know that if $m = 2$ or $n = 2$, (m, n) eventually reduces to one of the two repeating patterns $(**)$ or $(*)$. Now more generally, consider the case $(m, n) = (a, b)$ or (b, a), where $0 \leq a < b$. Then

$$\begin{matrix} (b, a) \to \\ (a, b) \to \end{matrix} (\underbrace{0, 0, \ldots, 0}_{a \text{ zeros}}, 1, \underbrace{0, 0, \ldots, 0}_{b-a-1 \text{ zeros}}, 1) \to (b - 1, 2)$$

which we have already shown eventually repeats one of the two patterns.

The only case remaining is $(m,n) = (k,k)$ for some $k \neq 2$. For small k, this is:

$$(0,0) \to (2) \to (0,0,1) \to (2,1)$$
$$(1,1) \to (0,2)$$
$$(3,3) \to (0,0,0,2) \to (3,0,1) \to (1,1,0,1) \to (1,3)$$

all three cases $(2,1), (0,2), (1,3)$ have already been covered. Now when $k \geq 4$,

$$(k,k) \to (\underbrace{0,0,\ldots,0}_{k \text{ zeros}}, 2) \to (k,0,1) \to (1,1,\underbrace{0,0,\ldots,0}_{k-2 \text{ zeros}}, 1) \to (k-2,3)$$

At this point either $k - 2 = 3$, or $k - 2 \neq 3$; we dealt with both of these cases before.

Therefore, the only pairs (m,n) which Joel could end up writing infinitely many times are those which appear in one of the two repeating patterns: $(1,2), (2,2)$, and $(0,3)$.

9. *(2022 P5, Solved By: 6%, Average Score: 0.8, Author: Caleb Stanford)*

We claim that the maximum possible value of $D(n)$ is 2. To start, we label the lily pads in clockwise order as pads 0, 1, 2, ..., 2021, such that the frog starts on pad 0. Additionally, if $n > 2021$, and $n \equiv r$ (mod 2022) for some $0 \leq r < 2022$, we may refer to "pad n" as another identifier for pad r.

We prove two lemmas.

Lemma 1: Suppose that $\gcd(k, 2022) = d$ (where gcd is the greatest common divisor function). Then the numbers $0, k, 2k, \ldots, (2022/d-1)k$ are all distinct modulo 2022.

Proof: If $j_1 k \equiv j_2 k$ (mod 2022), then $k(j_1 - j_2) \equiv 0$ (mod 2022), so $2022 \mid k(j_1 - j_2)$. Therefore, $(2022/d) \mid (k/d)(j_1 - j_2)$. But k/d and $2022/d$ are relatively prime (otherwise $\gcd(2022, k) > d$). Therefore, $(2022/d) \mid (j_1 - j_2)$. Since $0 \leq j_1, j_2 \leq 2022/d - 1$, it follows that $j_1 = j_2$. Therefore, it is impossible for the list to contain repeated elements. $\square$

Lemma 2: Suppose that $\gcd(k, 2022) = d$, the heights of lily pads 0, k, $2k$, ..., $(2022/d - 1)k$ are $k + 1$, and the heights of the remaining pads are k. Also, suppose that the frog is currently on pad 0. Let $\gcd(k + 1, 2022) = e$. Then after a series of jumps, the heights of lily pads 0, $(k + 1)$, $2(k + 1)$, ..., $(2022/e - 1)(k + 1)$ will all be $k + 2$, while the heights of the remaining pads will be $k + 1$, with the frog on pad 0.

Proof: At the start, we note that $0 \equiv k \equiv 2k \equiv \cdots \equiv (2022/d - 1)k \equiv 0$ (mod d), so there are $2022/d$ pads with height $k + 1$, each with pad numbers congruent to 0 (mod d). Since there are $2022/d$ pads with pad numbers congruent to 0 (mod d), this means that all of the pads congruent to 0 (mod d) have height $k + 1$.

If $d = 1$, this amounts to *all* of the pads having height $k + 1$. Therefore, the frog's next several jumps will be to pads $(k + 1)$, $2(k + 1)$, $3(k + 1)$, ..., until it reaches a pad of height $k + 2$. The pads with height $k + 2$ will be the pads it previously visited (i.e., pads 0, $(k + 1)$, $2(k + 1)$, ...). If $e = \gcd(k + 1, 2022)$, then all of these pads will be 0 (mod e), and by Lemma 1, we know that 0, $(k + 1)$, $2(k + 1)$, ..., $(2022/e - 1)(k + 1)$ will all be distinct modulo 2022. Since there are $2022/e$ pads that are divisible by e, we have covered each multiple of e exactly once. The next pad visited is $(2022/e)(k + 1) \equiv 0$ (mod 2022), so the frog returns to pad 0, having increased the heights of pads 0, $(k + 1)$, $2(k + 1)$, ..., $(2022/e - 1)(k + 1)$ to $k + 2$. This proves the conclusion of the lemma in this case.

Otherwise, if $d > 1$, then the frog first jumps to pad $k + 1$. This pad is congruent to 1 (mod d), so it has height k. This increases pad 0 to height $k + 2$. Since this pad [and all of the pads with pad numbers congruent to 1 (mod d)] has height k, its next jump will have length k, and as long as it jumps to distinct pads with pad numbers congruent to 1 (mod d), it will continue making jumps of length k. If all of its jumps have length k, then it visits the pads $(k+1)$, $(k+1)+k$, $(k+1)+2k$, ..., $(k + 1) + (2022/d - 1)k$. In particular, we note that these pad numbers are all congruent to 1 (mod d), and by Lemma 1, they must be distinct modulo 2022. Making one more jump, the frog will visit pad $(k + 1) + (2022/d)k \equiv k + 1$ (mod 2022) next, and at this point, all of the pads with numbers congruent to 1 (mod d) will have height $k + 1$.

This process continues—the frog jumps next to pad $2(k + 1)$, which is congruent to 2 (mod d), so it has height k. This increases pad $(k+1)$ to height $k+2$. A similar argument to the previous paragraph shows that all of the pads with numbers congruent to 2 (mod d) will have their heights increased to $k + 1$, and then pad $2(k + 1)$ will have its height increased to $k + 2$. This moves through all of the residue classes modulo d, and at the end of this process, pads 0, $(k + 1)$, $2(k + 1)$, ..., $(d - 1)(k + 1)$ will all have height $k + 2$, and the frog will be at pad $d(k + 1)$.

Now since $\gcd(k, k + 1) = 1$, where $d \mid k$ and $e \mid k + 1$, we know that $\gcd(d, e) = 1$. Therefore, d and e are relatively prime factors of 2022, so $de \leq 2022$. Hence $d \leq \frac{2022}{e}$, so $d - 1 \leq \frac{2022}{e} - 1$. In particular, the frog will keep jumping by $k + 1$ until it reaches pad $(2022/e - 1)(k + 1)$, at which point it jumps one more time (length $k + 1$) to arrive at pad $(2022/e)(k + 1) \equiv 0$ (mod 2022). This is the desired state, so the lemma is proven. □

At the start, all of the pads have height 1, so the frog jumps clockwise by 1 step, and it continues to do this until it returns to pad 0. At this point, each pad will have height 2. Note that this satisfies the conditions of Lemma 2 with $k = 1$.

Now if we have an arrangement of heights where the frog is at pad 0 and satisfies the initial conditions of Lemma 2, then the arrangement in the conclusion of Lemma 2 also satisfies the conditions of Lemma 2. Therefore, by induction, the frog's movements can be completely described by Lemma 2. In one application of Lemma 2, the shortest lily pad has height k, and the tallest lily pad has height $k + 2$, so $D(n)$ is less than or equal to $(k + 2) - k = 2$ at every point in time during the steps described by Lemma 2. Therefore, $D(n) \leq 2$. In particular, after 2022 steps, all of the lily pads have height 2. Then after 1011 additional steps, the heights of the lily pads are $3, 2, 3, 2, 3, 2, \ldots, 3, 2$ (where the frog is on pad 0). In the frog's next jump, the height of pad 0 increases to 4, so $D(3034) = 4 - 2 = 2$, hence the maximum value of $D(n)$ is 2.

3.2   COUNTING AND PROBABILITY

1. *(2018 Flyer, Author: Grant Molnar)*

   We will use inclusion-exclusion to show the answer is 141. First, the total number of pairs of points is the binomial coefficient 28 choose 2:

   $$\binom{28}{2} = 378.$$

   From these, we subtract pairs which lie on a line with three or more points. There are three types of such lines, based on the closest point to a point in the grid:

   (I) The closest point could be a single unit away, so the line is along one of the gridlines of the grid.

   (II) The closest point could be two gridlines away and one up, across a diamond shape.

   (III) The closest point could be three gridlines away and one or two units up.

   More than 3 gridlines away does not need to be considered because there are only 7 gridlines total, so there cannot be 3 points on the line if each is 4 gridlines away from the last.

   - For type (I) lines, there are, in one of three directions,

   $$\binom{7}{2} + \binom{6}{2} + \binom{5}{2} + \binom{4}{2} + \binom{3}{2} = 55,$$

   for a total of $3 \cdot 55 = 165$ pairs.

- For type (II) lines, there are, in one of three directions,

$$\binom{4}{2} + 4 \cdot \binom{3}{2} = 18,$$

for a total of $3 \cdot 18 = 54$ pairs.

- For type (III) lines, there are, in one of three directions,

$$2 \cdot \binom{3}{2} = 6,$$

for a total of $3 \cdot 6 = 18$ pairs.

Subtracting all the above cases, the total number of pairs of points left—and thus, the total number of lines passing through exactly two points—is

$$378 - 165 - 54 - 18 = 141.$$

Note: There was an error in the original version of this problem, published on the flyer: it displayed a grid with 36 dots (8 rows), instead of 28 dots (7 rows).

Note: As a function of the side length of the triangle n, the number of lines is given by the sequence (for $n = 1$ to $n = 15$)

$$0, 3, 6, 18, 39, 81, 141, 237, 369, 561, 801, 1119, 1521, 2043, 2667, \ldots$$

and is given by sequence A362014 in OEIS [12].

2. *(2017 P1, Solved By: 56%, Average Score: 4.7, Author: Caleb Stanford)*

There are 12 ways.

Solution 1: Since there are only 6 roads, and the cities cannot be fully connected by 2 or fewer roads, we must paint 3 roads blue and 3 orange. The number of ways to do this is $\binom{6}{3} = 20$.

Painting 3 roads of each color, it will always be possible to travel between any two cities on blue roads *unless* we paint the 3 blue roads in a triangle. Likewise for orange. It is impossible to have both a blue and an orange triangle. The number of possible triangles is $\binom{4}{3}$, and that triangle can be either orange or blue; this generates all incorrect colorings with 3 roads of each color.

Thus, the number of ways to color the roads correctly is

$$\binom{6}{3} - 2\binom{4}{3} = 20 - 8 = 12.$$

Solution 2: City 2's three roads cannot all be blue, or else it would not be connected by orange roads to the other cities. Similarly, its roads cannot all be orange. So it either has 2 blue and 1 orange, or 1 orange and 2 blue. There are 3 ways to have 2 blue and 1 orange, and 3 ways to have 2 orange and 1 blue, and by symmetry all these 6 cases are the same. So we can just count a particular case—assume $(1, 2)$ is orange and $(2, 3)$ and $(2, 4)$ are blue.

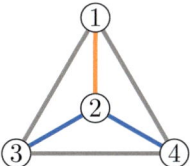

If both $(1, 3)$ and $(1, 4)$ are orange, 1 isn't connected by blue, and if both are blue, 3 and 4 aren't connected to 1 and 2 by orange. Thus, one of them is orange and one of them is blue; either way, $(3, 4)$ is forced to be orange and the resulting picture is a valid coloring. So there are 2 valid colorings.

Since there are 2 ways to complete the coloring in each of the 6 symmetric cases, the answer is $6 \times 2 = 12$.

3. *(2019 P1, Solved By: 43%, Average Score: 4, Author: Grant Molnar)*

There are 40 ways. We consider the following 3 cases: the largest square is 4×4; or the largest square is 3×3; or all squares are 1×1 or 2×2. In the last case, there are several subcases.

- If the largest square is a 4×4, then there is 1 way.
- If the largest square is a 3×3, then the square has 2 possible x-coordinates and 2 possible y-coordinates, so there are $2 \cdot 2 = 4$ ways.
- Otherwise, all squares are 2×2 or 1×1. We say that each 2×2 square is either a "corner square" (adjacent to a corner of the original 4×4 square), an "edge square" (adjacent to an edge of the original 4×4 square, but not adjacent to a corner), or a "center square" (not adjacent to an edge or a corner of the original 4×4 square).

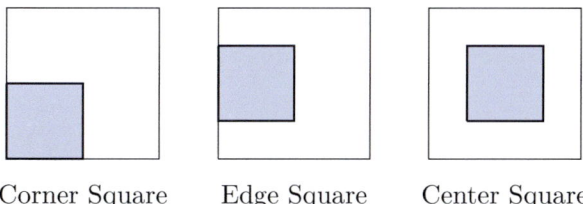

Corner Square   Edge Square   Center Square

There can be at most one center square and up to 2 edge squares. We proceed with subcases based on the presence of the center square and the number of edge squares:

> – If there is any center square, then no other 2×2 squares fit. So there is only 1 way.
>
> – Otherwise, if there are 2 edge squares, there are no corner squares and only 2 ways.
>
> – If there is 1 edge square, there are 4 rotations of the edge square and then two choices for whether to include a corner square in each of the opposite corners, for $4 \times (2 \times 2) = 16$ ways.
>
> – Finally, we arrive at the case where there are no center or edge squares. Then for each corner, we choose whether to put a corner square. So there are $2 \times 2 \times 2 \times 2 = 16$ ways.

Adding everything up, we get the answer:

$$1 + 4 + 1 + 2 + 16 + 16 = 40.$$

4. *(2020 P3, Solved By: 32%, Average Score: 2.8, Author: Hiram Golze)*

The required expected value is $128\,/\,63$.

Solution 1: The number of regions in the grid can be calculated in the following formula:

$$(\#\text{ regions}) = 4 - (\#\text{ adjacencies}) + (\#\text{ boxes}) \qquad (*)$$

where an *adjacency* is two filled squares that are horizontally or vertically adjacent, and a *box* is four filled squares in a 2×2 box. This formula is derived from the Euler characteristic for planar graphs. To see why it is true, we can apply the principle of inclusion-exclusion. First, we overcount the number of regions by counting the number of filled squares, which is 4. Next, whenever two filled squares are adjacent, that combines two regions into one; so we subtract 1 for each adjacency. However, there is a case where this might now undercount: if when combining two regions into one, those were already the same region. This happens only if there is a 2×2 box, where we have four filled squares, then four adjacencies, so that the final adjacency is between two regions that were already combined. To correct for this, we add 1 if there is a 2×2 box in the figure.

Now from equation $(*)$, we apply the principle of linearity of expectation to get:

$$\mathbb{E}[\#\text{ regions}] = 4 - \mathbb{E}[\#\text{ adjacencies}] + \mathbb{E}[\#\text{ boxes}].$$

(Here $\mathbb{E}[X]$ denotes the expected value of X; linearity of expectation states that for any X and Y, $\mathbb{E}[X + Y] = \mathbb{E}[X] + \mathbb{E}[Y]$.)

- To calculate $\mathbb{E}[\#\text{ adjacencies}]$, there are 12 possible adjacencies, and each adjacency happens with probability $\binom{7}{2}/\binom{9}{4}$, because there are $\binom{9}{4}$ ways to select four filled squares, and if two adjacent squares are filled in, the two remaining filled squares can be chosen in $\binom{7}{2}$ ways. So

$$\mathbb{E}[\text{\# adjacencies}] = 12 \cdot \frac{\binom{7}{2}}{\binom{9}{4}} = 12 \cdot \frac{21}{126} = 2.$$

- To calculate $\mathbb{E}[\text{\# boxes}]$, there are 4 ways to have a 2×2 box, so we have

$$\mathbb{E}[\text{\# boxes}] = \frac{4}{\binom{9}{4}} = \frac{4}{126} = \frac{2}{63}.$$

Therefore, our final answer is

$$4 - 2 + \frac{2}{63} = \frac{128}{63} \approx 2.032.$$

Solution 2: Alternatively, we can solve this using casework. In total, there are $\binom{9}{4} = 126$ ways to choose 4 squares out of 9. For the cases, we consider the number and sizes of the regions.

Case 1: One region, size 4.

There are various possible shapes here. For a 2×2 box, there are 4 ways. For an L-shape, 16 ways. For a T-shape, 8 ways. For a lightning bolt (zig-zag) shape, 8 ways. In total, 36 ways.

Case 2: Two regions, sizes 3 and 1.

If the size 3 region is a line, there are 12 ways. If the size 3 region is an L-shape, there are 32 ways (4 corners to put the L in, and 4 rotations where the rotations have 1 way, 2 ways, 2 ways, and 3 ways). In total, there are 44 ways in this case.

Case 3: Two regions, sizes 2 and 2.

Here the center square cannot be filled. There are two corners and two edges filled. There are $2 \cdot 4 = 8$ ways if the corners are opposite, and 4 ways if the corners are adjacent, for 12 ways total.

Case 4: Three regions, sizes 2, 1, and 1.

If the center square is filled there are 4 ways. If not, there are 8 ways to place the region of size 2, then 3 ways for the remaining two regions, for 24 ways total. So 28 ways in this case total.

Case 5: Four regions, sizes 1, 1, 1, 1.

If the center is filled there are 4 ways; otherwise 2 ways, for 6 total.

We can also double check that we did the casework correctly by adding up the total number of ways above. We get $36 + 44 + 12 + 28 + 6 = 126 = \binom{9}{4}$, as expected.

Now to finish the problem, we compute the expected value:

$$\frac{(36) \cdot 1 + (44 + 12) \cdot 2 + (28) \cdot 3 + (6) \cdot 4}{126} = \frac{256}{126} = \frac{128}{63}.$$

Note: If each square is shaded or unshaded independently at random (instead of choosing exactly 4 squares), then one can show that the expected number of regions is instead

$$1 + \frac{3}{4} + \frac{1}{512} = \frac{897}{512} \approx 1.752.$$

5. *(2016 P2, Solved By: 30%, Average Score: 2.8, Author: Grant Molnar)*

The required probability is $7\,/\,27$.

Solution 1: We split this into cases based on the form of the unordered values shown on the dice. In the following, $1 \le a, b, c, d \le 6$ are distinct numbers. After determining the (unordered) numbers that can appear on the dice, we must determine how many ways we can order those dice rolls in sequence.

Case 1: The unordered numbers shown are $\{a, a, a, a\}$.

In this case, it is clear that the pairing (a, a) and (a, a) will yield equal sums. We have 6 possible choices for a, and thus 6 possible rolls that take this form.

Case 2: The unordered numbers shown are $\{a, a, a, b\}$.

In this case, one of the pairs will contain b, hence the pairs will be (a, b) and (a, a). But these can never be equal as $a \ne b$.

Case 3: The unordered numbers shown are $\{a, a, b, b\}$.

To obtain equal sums, we must split the numbers into the pairs (a, b) and (a, b). There are $\binom{6}{2} = 15$ ways to select a and b, and there are $\binom{4}{2} = 6$ ways to order the rolls (select two of the rolls to be a's). Hence there are $15 \cdot 6 = 90$ total rolls that take this form.

Case 4: The unordered numbers shown are $\{a, a, b, c\}$.

In this case, either b pairs with a or b pairs with c. If b pairs with a, then $a + b = c + a$, which is impossible as $b \ne c$. Therefore, b must pair with c, so $2a = b + c$. Therefore, $b + c$ is even. We may assume without loss of generality that $b < c$, hence the possibilities for (b, c) are

$$(b, c) = (1, 3), (1, 5), (2, 4), (2, 6), (3, 5), (4, 6).$$

Therefore, there are six ways to choose (b, c), and a is automatically determined (it is the average of b and c). Then there are $\frac{4!}{2!} = 12$ ways to order the rolls, hence there are $6 \cdot 12 = 72$ total rolls that take this form.

Case 5: The unordered numbers shown are $\{a, b, c, d\}$.

In this case, the sum of the pairs must be representable in at two ways with distinct integers. We find

$$5 = 1 + 4 = 2 + 3$$
$$6 = 1 + 5 = 2 + 4$$
$$7 = 1 + 6 = 2 + 5 = 3 + 4$$
$$8 = 2 + 6 = 3 + 5$$
$$9 = 3 + 6 = 4 + 5.$$

These are the only numbers that can be represented as the sum of two numbers in at least two ways using distinct numbers. In particular, for the numbers $5, 6, 8, 9$, we know immediately what the four numbers a, b, c, d are. For 7, we must choose two of the three pairs, and we can do this in 3 ways. Therefore, the number of ways to choose $\{a, b, c, d\}$ is $1 + 1 + 3 + 1 + 1 = 7$. Then, we can order the rolls in $4! = 24$ ways. Thus there are $7 \cdot 24 = 168$ total rolls that take this form.

Adding these together, we find $6 + 90 + 72 + 168 = 336$ possibilities, so the answer is

$$\frac{336}{6^4} = \frac{7}{27}.$$

Solution 2: Let the four numbers be (a, b, c, d). Consider these three events:

   I. $a + b = c + d$

  II. $a + c = b + d$

 III. $a + d = b + c$.

We want the probability of "I or II or III." By the principle of inclusion-exclusion, this is

$$\mathbb{P}[\text{I}] + \mathbb{P}[\text{II}] + \mathbb{P}[\text{III}]$$
$$-\mathbb{P}[\text{I and II}] - \mathbb{P}[\text{I and III}] - \mathbb{P}[\text{II and III}]$$
$$+\mathbb{P}[\text{I and II and III}].$$

(Here, $\mathbb{P}$ means the probability of an event.) Symmetry in the problem implies that I, II, and III have equal probability, as well as "I and II," "I and III," and "II and III." Therefore, the probability we are looking for is simply

$$3\mathbb{P}[\text{I}] - 3\mathbb{P}[\text{I and II}] + \mathbb{P}[\text{I and II and III}]. \tag{1}$$

To compute $\mathbb{P}[1]$, split into cases based on the value of the sum $s = a + b$. We have

$$\mathbb{P}[1] = \mathbb{P}[a + b = c + d] = \sum_{s=2}^{12} \mathbb{P}[a + b = c + d = s]$$

$$= \sum_{s=2}^{12} \mathbb{P}[a + b = s] \cdot \mathbb{P}[c + d = s]$$

which expands to

$$\frac{1^2 + 2^2 + 3^2 + 4^2 + 5^2 + 6^2 + 5^2 + 4^2 + 3^2 + 2^2 + 1^2}{6^4} = \frac{146}{6^4}.$$

The other two are easier. If I and II both hold, then $b = c$ and $a = d$, and if $b = c$ and $a = d$, then I and II both hold. The probability of this is simply

$$\mathbb{P}[\text{I and II}] = \mathbb{P}[a = d \text{ and } b = c] = \mathbb{P}[a = d] \cdot \mathbb{P}[b = c] = \frac{1}{6^2}.$$

Finally, if I and II and III all hold, that means $a = b = c = d$, which can happen in exactly 6 ways, so

$$\mathbb{P}[\text{I and II and III}] = \mathbb{P}[a = b = c = d] = \frac{1}{6^3}.$$

From (1), the desired probability is thus

$$3 \cdot \frac{146}{6^4} - 3 \cdot \frac{1}{6^2} + \frac{1}{6^3} = \frac{73 - 18 + 1}{6^3} = \frac{56}{6^3} = \frac{7}{27}.$$

6. (*2015 P5, Solved By: 18%, Average Score: 2.1, Author: Caleb Stanford*)

There are 2016 ways. To show this, imagine each number in the grid is written using its prime factorization. These factorizations consist of products of 2 and 5, and possibly factors of -1 to account for negative integers. We will fill the grid by placing into the grid all the factors of 5, 2, and -1 in turn. Since prime factorization is unique, this is equivalent to the original problem. (This strategy is known as a *bijection proof*.)

Beginning with the factors of 5, there are 3 ways to place a 5 in the first column, 2 remaining possible slots in the second column, and one remaining possible slot in the third column, for a total of $3 \cdot 2 \cdot 1 = 6$ ways to distribute 5s in the grid.

For the factors of 2, we can place factors of either 2^1 or $2^2 = 4$ in individual entries of the grid. We split into cases:

Case 1: The grid contains at least two factors of 4s.

Then the second 4 must be placed in a different row and column in the first, and this leaves only one row and column left which is forced to contain a third 4. So this subcase is essentially the same as the factors of 5, and it can be done in 6 ways.

Case 2: The grid contains exactly one factor of 4.

Then the entries in the same row/column as the 4 must be 1s, and the other entries must all be 2s. Everything is determined by the location of the factor of 4, so this can be done in 9 ways.

Case 3: The grid contains no 4s.

Then each row/column contains two 2s and one 1. Therefore, each row/column contains exactly one 1. Therefore, this can be counted in exactly the same method as the 5s above, for a total of 6 ways.

Adding up the cases, we can place the 2s in a total of $6 + 9 + 6 = 21$ ways.

Lastly, for the factors of -1, any particular distribution of signs in the grid is uniquely determined by the sign distribution in the upper-left-hand 2×2 grid of boxes, since this forces the signs in the third row and third column. Moreover, every particular choice of signs in the 2×2 grid is consistent. Therefore, the factors of -1 can be distributed in $2^4 = 16$ ways.

Putting everything together, by placing the factors of 5, 2, and -1 in turn, the total number of ways of filling the grid is $6 \times 21 \times 16 = 2016$.

7. *(2014 P6, Solved By: 27%, Average Score: 1.7, Author: Hiram Golze)*

First, we classify the steps into two categories: *forward* steps and *lateral* steps, as labeled below.

Forward Steps   Lateral Steps

In a path from A to B, there are exactly $2n$ forward steps. To see this, split the diagram at what we will term *levels* as shown in the following picture. Each level consists of the segments along which we can move laterally without moving forward.

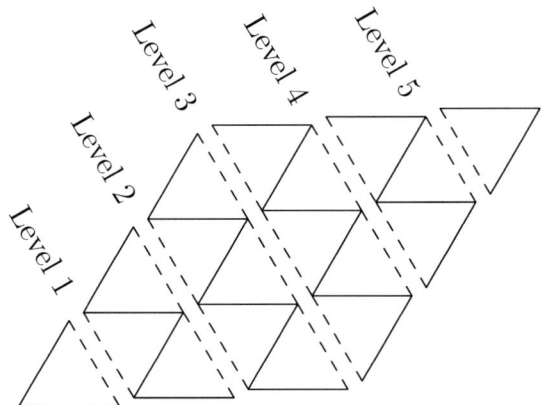

A path consists of a step from Level 0 to level 1 (the first forward step), followed by some sequence of steps in level 1, followed by a step from level 1 to level 2 (the second forward step), followed by some sequence of steps in level 2, and so on. Note that due to the restrictions on steps, once we reach a level, we cannot return to a previous level. Therefore, we can only make one forward step between levels, so we will make exactly $2n$ forward steps.

Given a path, we split it up into its forward steps and its lateral steps. We claim that the lateral steps are completely determined by our choice of forward steps. This is shown by the following bijection: if we know the forward steps of a path, then the nth forward step will start at level $n-1$ and end at a point on level n. The $(n+1)$th forward step will start at another point on level n. But then there is a unique path in level n between these points without retracings because we can only move in one direction in a level. Therefore, every path is uniquely determined by its forward steps.

Also, given a sequence of $2n$ forward steps with one forward step in each gap between levels, we can fill in lateral steps uniquely as well, and hence there is a unique path associated to each sequence of $2n$ forward steps. This establishes a one-to-one correspondence (or bijection) between paths and sequences of $2n$ forward steps, each starting at a different level.

Therefore, the number of paths is equal to the number of ways to select $2n$ forward steps, with exactly one forward step between two levels. Referring to the above diagram, for the first forward step, we have 2 choices, for the second forward step we have 4 choices, ..., for the nth forward step we have $2n$ choices, for the $(n+1)$th forward step we have $2n$ choices, for the $(n+2)$th forward step we have $2(n-1)$ choices, ..., for the $(2n)$th forward step we have 2 choices. Therefore, the total number of paths is

$$(2 \cdot 4 \cdot 6 \cdots (2n))((2n) \cdot (2n - 2) \cdots 2) = (2 \cdot 4 \cdot 6 \cdots (2n))^2.$$

This is always a perfect square, so the result follows.

Note: More concisely, the answer can be written as

$$(2^n n!)^2 = 4^n (n!)^2.$$

These numbers are sometimes known as the *central factorial numbers*, and are given by sequence A002454 in OEIS [8].

8. *(2018 P5, Solved By: 12%, Average Score: 1, Author: Caleb Stanford)*

Solution 1: We show that the number of ways is $12 \cdot 11 \cdot 10 = 1320$. In fact, this generalizes. Suppose we want to color n points around a circle with $a_1, a_2, a_3, \ldots, a_k$ of each color, where k is the number of colors and $a_i \geq 1$ are fixed, satisfying the rule that each pair of colors is separated by a line. Then the number of ways is given by the *falling factorial*

$$n(n - 1)(n - 2) \cdots (n - k + 1).$$

While the argument presented here works in the general case, we focus on the specific case of 4 colors with $3, 3, 3$, and 3 of each color.

To prove this, we will use a bijection proof. Consider the following procedure for coloring the circle. We start out with three disks each of red, green and blue. For each of these three colors (excluding white), we pick a single point (called the source) where we place the stack of three disks of that color. Thus we have chosen three distinct points—a red source, a green source, and a blue source. Now, we color the circle as follows.

- Start at point 1 and move around the points of the circle clockwise. As we move around the circle, we will be picking up and dropping off disks, so we will hold a vertical *stack* of disks in our hands (initially the stack is empty). As we visit each successive point, there are several possible scenarios.
  - If the point that we visit is a source that we have not visited yet, we pick up the three disks of that color and add those disks to the *top* of our stack. However, we also immediately place the top disk from our stack on that point.
  - If the point that we visit does not have any disks placed on it, and the stack of disks in our hand is empty, we move on to the next point.
  - If the point that we visit does not have any disks placed on it, and the stack of disks in our hand is not empty, we place the top disk from our stack on that point.

– If the point that we visit already has a disk placed on it, move on to the next point.

• We keep going around the points of the circle in clockwise order until we have visited each source, and then we continue going around the circle clockwise until our stack is gone.

Since there are 9 disks and 12 points, this process will definitely stop. When it stops, we color the three points without a disk white.

To illustrate this process, consider the following example. Suppose we initially place a red stack at point 2, a blue stack at point 4, and a green stack at point 6. As we move around the circle clockwise from point 1, our steps are demonstrated in the table below. The first row shows the stack of disks in our hand before visiting the point. The second row (+) shows disks picked up from the point, and the third row (−) shows disks dropped at the point. Finally, the last row shows disks after visiting the point.

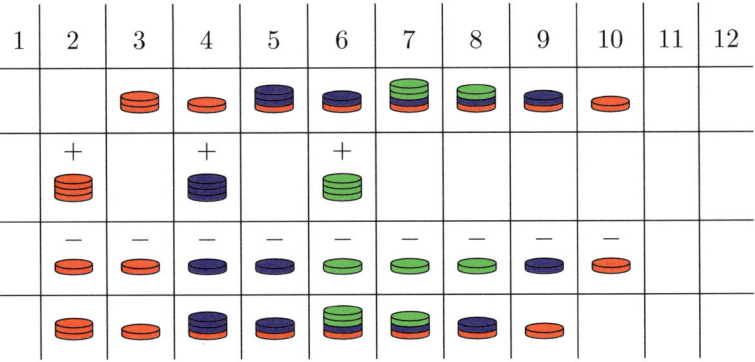

Therefore, we can draw the following two diagrams, illustrating the initial placement of the stacks (left), and the final coloring that results (right).

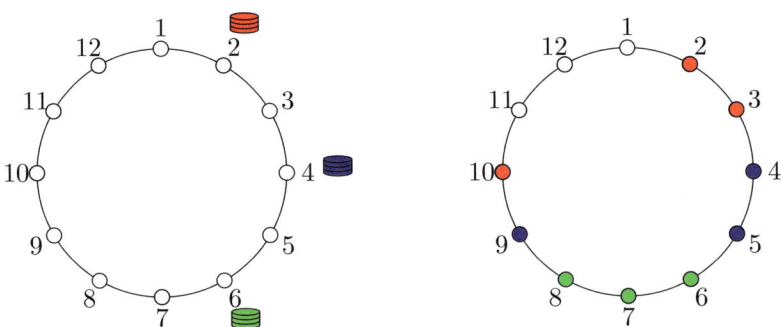

We first show this coloring is valid. As we use three disks of each color for red, green, and blue disks, that leaves three points colored white, so we color three points of each color. Because of the stacking method, we never drop a red disk, then a blue disk, then a red disk, then a blue disk again; once we pick up the blue stack, we don't drop a red disk again again until the blue disks are gone. Also, we only color points with white on points where, every time crossing that point, we were holding no disks, so since we pick up a stack of a single color and drop all disks before we hold no disks again, we can draw a line that separates the disks of that color from the white points. Therefore, each pair of colors can be separated by a line.

Finally we have to show that this painting procedure generates every possible coloring exactly once. Starting from a coloring of the circle, the white dots separate the circle into three sections. Note that points of the same color can be in at most one of these sections, so for each color, we place a stack of three disks of that color on the first point in that section (in clockwise order) that is of that color. For example, in the above example, we split the circle into sections as shown below.

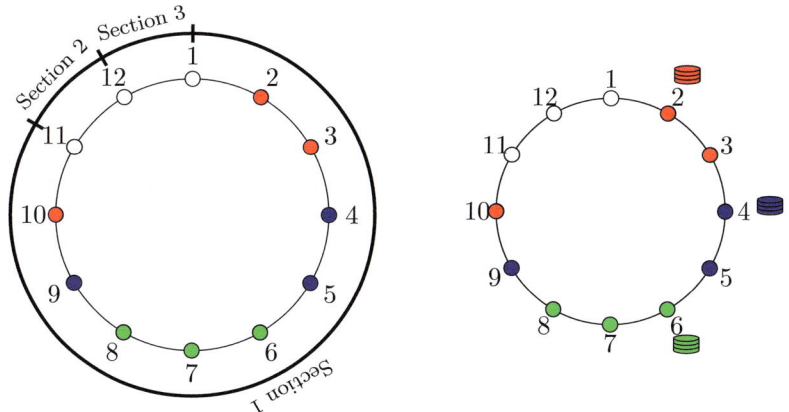

Since point 2 is the first red point in Section 1, we place the red stack at point 2. Since point 4 is the first blue point in Section 1, we place the blue stack at point 4. Since point 6 is the first green point in Section 1, we place the green stack at point 6.

If we do this and then do the previous procedure, we obtain the original coloring, and conversely, if we do the previous procedure and then this, we recover the original locations of the stacks. Therefore, the number of total colorings is the same as the number of ways to place the stacks.

Solution 2: Suppose that we have a valid coloring of the 12 dots. Then we draw all segments between points of the same colors. This creates four triangles, which we call the red, blue, green, and white triangles, and we shade each triangle with their color as shown below.

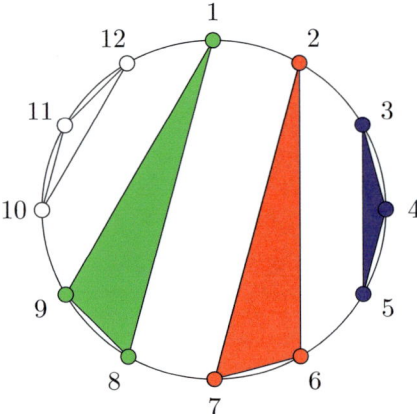

We first argue that a coloring is valid if and only if the four colored triangles do not overlap. If the coloring is valid, then consider for example the line separating the red and white points; then this line also separates the red and the white triangle. Conversely, if the red and white triangles do not overlap, then one of the three edges of the red triangle separates the red and white points (if it is moved a small amount away from the red triangle).

The 12 points split the circle into 12 equally-sized arcs. Call a triangle an (a, b, c) triangle if, when reading around the circle clockwise, the distance from the first point to the second is a, the distance from the second to the third is b, and the distance from the third back to the first is c (in number of arcs). For example, the blue triangle is a $(1, 1, 10)$-triangle. We claim that for each triangle, if it is an (a, b, c)-triangle, then $a+b+c = 12$ and $a, b, c \in \{1, 4, 7, 10\}$. The fact $a + b + c = 12$ is because the total distance around the circle is 12. For the second fact, note that since no two triangles intersect, each of the remaining three triangles must have all of its points on one of the three arcs that the triangle splits the circle into. Therefore, the number of points between two consecutive points on the initial triangle is always a multiple of 3, which means that the number of equally-sized arcs between those two consecutive points is one more than a multiple of 3. Hence $a, b, c \in \{1, 4, 7, 10\}$.

It follows that every triangle is either $(1, 4, 7)$, $(4, 4, 4)$, or $(1, 1, 10)$, up to reordering the triple. We proceed with casework based on which triples appear.

Case 1: One of the triangles is $(4, 4, 4)$.

There are 4 choices for the color of the triangle with triple $(4, 4, 4)$, and 4 rotations at which the $(4, 4, 4)$ triangle can be placed. Once the $(4, 4, 4)$ triangle is placed, the positions of the other three triangles are forced, so we simply have 3! ways to color the remaining triangles. Thus there are $4 \cdot 4 \cdot 3! = 96$ such colorings.

Case 2: One of the triangles is $(1, 4, 7)$.

The side of length 4 arcs forces a single $(1, 1, 10)$-triangle to be placed, whose position is forced. However, for the side of length 7 arcs, we have two subcases:

Subcase 2.1: The other two triangles are $(1, 1, 10)$ and $(1, 1, 10)$. In this case, there are 4 choices for the color of the $(1, 4, 7)$ triangle, and 24 choices for the position of the $(1, 4, 7)$ triangle (12 choices for the point between the arcs of length 4 and 7, and then 2 reflections). Since the remaining triangles are all $(1, 1, 10)$ triangles, their positions are uniquely determined, and there are 3! choices for their colors. Thus there are $24 \cdot 4 \cdot 3! = 576$ colorings here.

Subcase 2.2: The other two triangles are $(1, 4, 7)$ and $(1, 1, 10)$. In this case, note that the two $(1, 4, 7)$ triangles must have the sides of length 7 parallel and next to each other on either side of a diameter of the circle. There are 6 choices for this diameter. But we still must place the final point in each of these triangles, and due to reflections (perpendicular to the chosen axis), there are $2 \cdot 2$ choices for the third points of these triangles. Also, there are $4 \cdot 3$ choices for the colors of the two $(1, 4, 7)$ triangles. The positions of the final two triangles are uniquely determined, but there are 2 choices for the colors of the remaining triangles. Thus there are $6 \cdot (2 \cdot 2) \cdot (4 \cdot 3) \cdot 2 = 576$ colorings here.

Case 3: All the triangles are $(1, 1, 10)$ triangles.

In this case, the relative positions of the triangles are determined, but there are 3 distinct rotations. Then there are 4! ways to color the triangles, for a total of $3 \cdot 4! = 72$ such colorings.

Therefore, there are a total of $96 + 576 + 576 + 72 = 1320$ colorings.

9. *(2021 P4, Solved By: 5%, Average Score: 0.7, Author: Daniel South)*

The answer is that $P_c(2021)$ is a multiple of 3 if and only if $c \equiv 2 \pmod 3$.

Solution 1: First, scale the problem so that cows only have two legs and ostriches have one, and the number of animals on day n is equal to the number of legs on day $n - 1$. Then, the number of animals Georgia adds on day n is simply the number of cows on day $n - 1$, since they have one excess leg.

The possible numbers of cows on day 2 are $c, c+1, c+2, \ldots, 2c$. Thus, we have the recurrence relation

$$P_c(n) = \sum_{i=c}^{2c} P_i(n-1)$$

for $n \geq 2$, with the initial condition $P_c(1) = 1$ for all c.

We now claim that for all $n \geq 2$ we have $P_c(n) \equiv c+1 \pmod 3$. We prove this by induction. The base case is day 2, on which Georgia can add $0, 1, 2, \ldots c$ cows, which gives $c+1$ possibilities. Now suppose that $P_c(n) \equiv c+1 \pmod 3$ for all c; we show that $P_c(n+1) \equiv c+1 \mod 3$ as well.

Working modulo 3, we have

$$P_c(n+1) \equiv \sum_{i=c}^{2c} (i+1)$$

$$\equiv c+1 + \sum_{i=c}^{2c} i$$

$$\equiv c+1 + \frac{2c(2c+1)}{2} - \frac{c(c-1)}{2}$$

$$\equiv c+1 + \frac{3c^2 + 3c}{2}$$

$$\equiv c+1 \pmod 3.$$

This completes the induction. It follows that $P_c(2021) \equiv 0 \pmod 3$ if and only if $c \equiv 2 \pmod 3$.

Solution 2: Extend $P_c(n)$ to the case $c = 0$ for convenience (in this case there are 0 cows always, so $P_c(n) = 1$ for all n). We first observe that for $c \geq 0$, $n \geq 1$, $P_c(n)$ is equivalently the number of sequences $c_1, c_2, \ldots, c_n$ where $c_1 = 1$ and $c_{i+1} \in [c_i, 2c_i]$ for all i. Here c_i is the number of cows on day i, and the number of ostriches on day i is derived as $2c_{i-1} - c_i$.

Then for $c \geq 1$ and $n \geq 1$, we have the recurrence

$$P_c(n+1) = P_{c-1}(n+1) - P_{c-1}(n) + P_{2c-1}(n) + P_{2c}(n),$$

by the following bijection: take a sequence starting with c and subtract 1 from the first number. Either it is a valid sequence starting with $c-1$, or the second number is $2c-1$, or the second number is $2c$. In the first case, this covers all valid sequence starting with $c-1$ other than those where the second number is $c-1$, so the number of such sequences is $P_{c-1}(n+1) - P_{c-1}(n)$.

From the above recurrence, we now claim that $P_c(n) \equiv c + 1 \pmod 3$ for all $n \geq 2$. (For $n = 1$, $P_c(1) = 1$ so it doesn't hold.) Induct on n. For the base case, $P_c(2)$ counting valid sequences of length 2, of which there are exactly $c + 1$. For the inductive step, using the recurrence:

$$P_c(n+1) \equiv P_{c-1}(n+1) - (c) + (2c) + (2c+1) \equiv P_{c-1}(n+1) + 1, \pmod 3$$

and since $P_0(n + 1) = 1$, the result follows.

10. *(2022 P6, Solved By: 0%, Average Score: 0.3, Author: Caleb Stanford)*

To begin, let x_n be the number of fish-friendly $2 \times n$ grids. Let a_n be the number of fish-friendly $2 \times n$ grids where the two squares in the last column are blue, and let b_n be the number of fish-friendly $2 \times n$ grids where only one square in the last column is blue. Since at least one square in the last column of a fish-friendly grid is blue, we find

$$x_n = a_n + b_n.$$

To compute a_n, note that if a fish-friendly $2 \times n$ has both squares colored blue in the last column, then the prior $2 \times (n-1)$ grid can be any fish-friendly grid, hence

$$a_n = a_{n-1} + b_{n-1}. \tag{1}$$

To compute b_n, consider the second-to-last column. If it has both squares colored blue, then the first $2 \times (n-2)$ grid can be any fish-friendly grid and there are two choices for the last column, so there are $2a_{n-1}$ such grids. If it has only one square colored blue, then the last column is forced to be the same as the second-to-last column, and there are b_{n-1} such grids. Thus,

$$b_n = 2a_{n-1} + b_{n-1}. \tag{2}$$

Adding (1) and (2), we find

$$a_n + b_n = 3a_{n-1} + 2b_{n-1}.$$

Hence $x_n = 2x_{n-1} + a_{n-1}$. Since $a_{n-1} = a_{n-2} + b_{n-2} = x_{n-2}$, we deduce that

$$x_n = 2x_{n-1} + x_{n-2},$$

where $x_1 = 3$ and $x_2 = 7$. This is a homogeneous linear recurrence with characteristic equation $\lambda^2 - 2\lambda - 1 = 0$, which has roots $\lambda_1 = 1 + \sqrt 2$ and $\lambda_2 = 1 - \sqrt 2$. By the theory of homogeneous linear recurrences, there exist constants c_1 and c_2 such that $x_n = c_1(1 + \sqrt 2)^n + c_2(1 - \sqrt 2)^n$. Using the initial conditions $x_1 = 3$ and $x_2 = 7$, we find $3 = c_1(1 + \sqrt 2) + c_2(1 - \sqrt 2)$ and $7 = c_1(1 + \sqrt 2)^2 + c_2(1 - \sqrt 2)^2$. Multiplying the first equation by $(1 - \sqrt 2)$ and subtracting it from the second equation, we find $4 + 3\sqrt 2 = (4 + 2\sqrt 2) \cdot c_1$. Multiplying by $2 - \sqrt 2$, we find

$2 + 2\sqrt{2} = 4c_1$, so $c_1 = \frac{1+\sqrt{2}}{2}$. We plug this into the equation to find $c_2 = \frac{1-\sqrt{2}}{2}$. Therefore, the solution to this recurrence is

$$x_n = \frac{1}{2}\left((1+\sqrt{2})^{n+1} + (1-\sqrt{2})^{n+1}\right).$$

It follows that

$$x_{49} = \frac{1}{2}\left((1+\sqrt{2})^{50} + (1-\sqrt{2})^{50}\right).$$

The term $(1-\sqrt{2})^{50}$ is very small, so x_{49} can be approximated by the first term. In particular, $(1-\sqrt{2})^{50} > 0$, so

$$x_{49} > \frac{1}{2} \cdot (1+\sqrt{2})^{50}.$$

Now $(1+\sqrt{2})^2 = 3 + 2\sqrt{2}$, so $(1+\sqrt{2})^4 = 17 + 12\sqrt{2}$, and $(1+\sqrt{2})^5 = 41 + 29\sqrt{2}$. Hence $(1+\sqrt{2})^{10} = 3363 + 2378\sqrt{2}$. Then $3363 + 2378\sqrt{2} > 1.6 \cdot 2^{12}$. Therefore,

$$(1+\sqrt{2})^{50} > (1.6 \cdot 2^{12})^5 = 1.6^5 \cdot 2^{60}.$$

Since $1.6^5 = 10.48576 > 2^3$, we find that $(1+\sqrt{2})^{50} > 2^3 \cdot 2^{60} = 2^{63}$. Hence

$$x_{49} > \frac{1}{2} \cdot 2^{63} = 2^{62}.$$

Now suppose that we cover the top two rows of the 42×49 grid with a fish-friendly 2×49 grid, which we can do in at least 2^{62} ways. Then we can color the rest of the grid in $2^{40 \cdot 49} = 2^{1960}$ ways, since each of the remaining squares has two choices of color. Hence there are at least $2^{62+1960} = 2^{2022}$ fish-friendly colorings.

Note 1: It is possible to improve this count, which would give some leeway on computing the lower bound of x_{49}. We can cut the 42×49 grid into 21 distinct 2×49 grids, denoted $G_1, G_2, \ldots, G_{21}$. Let A_i denote the set of all colorings of the 42×49 grid such that G_i is fish-friendly. Observe that the number of fish-friendly colorings is certainly greater than or equal to $|A_1 \cup A_2 \cup \cdots \cup A_{21}|$. We use the fact that

$$\begin{aligned}
|A_1 \cup A_2 \cup \cdots \cup A_{21}| \geq &(|A_1| + |A_2| + \cdots + |A_{21}|) \\
&- (|A_1 \cap A_2| + |A_1 \cap A_3| \\
&+ \cdots \\
&+ |A_{20} \cap A_{21}|. \quad (3)
\end{aligned}$$

This is related to the principle of inclusion-exclusion. If a coloring is in exactly k of the sets A_i, then it is counted k times in the first set of parentheses, and it is subtracted $\binom{k}{2}$ times in the second set of parentheses. In particular, if $k = 0$, the coloring is counted 0 times, if $k = 1$, the coloring is counted once, if $k = 2$, the coloring is counted $2 - \binom{2}{2} = 1$ time, and if $k \geq 3$, then $k - \binom{k}{3} \leq 0$. So each coloring in $A_1 \cup A_2 \cup \cdots \cup A_{21}$ is counted at most once by the right-hand side of (3), so the left-hand side is greater than or equal to the right-hand side.

Using the same methods as above, there are $x_{49} \cdot 2^{40 \cdot 49}$ fish-friendly colorings in A_i. Also, $|A_i \cap A_j| = x_{49}^2 \cdot 2^{38 \cdot 49}$. Since there are $\binom{21}{2}$ intersections, we can use (3) to find that

$$|A_1 \cup A_2 \cup \cdots \cup A_{21}| \geq 21 \cdot x_{49} \cdot 2^{1960} - \binom{21}{2} \cdot x_{49}^2 \cdot 2^{1862}$$

$$= 2^{1862} \cdot 21 \cdot x_{49} \cdot (2^{98} - 10x_{49}).$$

Also, $x_{49} < \frac{1}{2}((1 + \sqrt{2})^{50} + 1)$, and $(1 + \sqrt{2})^5 = 41 + 29\sqrt{2} < 2^7$, so $x_{49} < \frac{1}{2}(2^{70} + 1) < 2^{70}$. Hence $10x_{49} < 16 \cdot 2^{70} < 2^{74}$, so $2^{98} - 10x_{49} > 2^{98} - 2^{74} > 2^{97}$. Hence the number of fish-friendly colorings is at least

$$2^{1862} \cdot 21 \cdot x_{49} \cdot (2^{98} - 10x_{49}) > 2^{1862} \cdot 21 \cdot 2^{62} \cdot 2^{97}$$

$$> 2^{1862+4+62+97} = 2^{2025}.$$

Note 2: In fact, using a computer, we can show that the number of fish-friendly colorings is much larger. By randomly generating grids and testing if they are fish-friendly, we estimate that approximately 0.086% of grids are fish-friendly (out of 100 million trials). Since there are $2^{42 \times 49} = 2^{2058}$ grids total, the number of fish-friendly grids is on the order of $.00086 \times 2^{2058}$, or about 2^{2048}.

This is only an estimate, however. We can also use a computer to get a *strict* lower bound of at least 2^{2043} fish-friendly grids. In particular, if the grid is split into four 10×49 subgrids and a 2×49 subgrid, the number of colorings that contain a fish-friendly path in one of these five subgrids that never goes backward is at least $2^{2043.696}$, using the same inclusion-exclusion argument as in Note 1.

Note 3: The number of fish-friendly $m \times n$ grids is given by the following table (see sequences A359576 and A365988 in OEIS [11, 13]):

1	1	1	1	1	...
3	7	17	41	99	...
7	37	197	1041	5503	...
15	175	1985	22193	247759	...
31	781	18621	433809	10056959	...
...					

11. *(2019 P6, Solved By: 0%, Average Score: 0.3, Author: Caleb Stanford)*

Solution 1: Let the n balls be of k different colors, and let there be a_i balls of each color i for $1 \leq i \leq k$. So $\sum_{i=1}^{k} a_i = n$. Then we would like to have $\sum_{i=1}^{k} \binom{a_i}{2} = \frac{1}{2}\binom{n}{2}$, i.e., the number of ways to pick two balls of the same color is half the total number of ways to pick two balls. However, if any a_i are 1 they contribute nothing to the sum. So it suffices to prove that for any $n \geq 200$, $n \equiv 0$ or $1 \pmod 4$, we can choose nonnegative integers a_i such that

$$\frac{1}{2}\binom{n}{2} = \sum_{i=1}^{k}\binom{a_i}{2} \quad \text{and} \quad \sum_{i=1}^{k} a_i \leq n.$$

Define a *monochromatic pair* to be a pair of balls with the same color. Thinking in terms of limited resources, the problem is to spend *at most* n balls to create *exactly* $\frac{1}{2}\binom{n}{2}$ monochromatic pairs, by distributing the balls into different colors.

First, pick a maximally so that $\binom{a}{2} \leq \frac{1}{2}\binom{n}{2}$, and color a balls red. We claim that this leaves us with *at least* $\frac{1}{4}n$ balls left to color, and *at most* $\frac{3}{4}n$ monochromatic pairs left to create. To show this, note that $\frac{1}{2}(a-1)^2 \leq \binom{a}{2} \leq \frac{1}{2}\binom{n}{2} \leq \frac{1}{4}n^2$. Rearranging, $a \leq 1 + \frac{1}{\sqrt{2}}n$. This is less than $\frac{3}{4}n$ as long as n is sufficiently large; in particular, as long as $\left(\frac{3}{4} - \frac{1}{\sqrt{2}}\right) n > 1$, which is true for $n \geq 25$ since

$$\left(\frac{3}{4} - \frac{1}{\sqrt{2}}\right) n > (.75 - .71)\, n = \frac{n}{25} \geq 1.$$

So we colored at most $a \leq \frac{3}{4}n$ balls. Additionally, since a was maximal, $\binom{a}{2} + a = \binom{a+1}{2} > \frac{1}{2}\binom{n}{2}$. Hence the number of monochromatic pairs left to create, or $\frac{1}{2}\binom{n}{2} - \binom{a}{2}$, is less than a, and thus less than $\frac{3}{4}n$.

Therefore, after coloring these a balls red, we are left with at least $\frac{1}{4}n$ balls to color in order to create at most $\frac{3}{4}n$ monochromatic pairs. At this point, we color groups of 13 balls at a time with their own color. Coloring 13 balls with the same color creates $\binom{13}{2} = 78$ monochromatic pairs. Therefore, if we use $b \leq \left\lfloor \frac{3n/4}{78} \right\rfloor$ different groups of 13 balls (where $\lfloor \cdot \rfloor$ is the floor function), where each group is monochromatic, and where b is chosen maximally, then we have at most 77 monochromatic pairs left to create, and we also colored at most $13 \cdot \frac{3n/4}{78} = \frac{n}{8}$ balls. We are left with at least $\frac{1}{4}n - \frac{1}{8}n = \frac{1}{8}n$ balls, and at most 77 monochromatic pairs to go.

To create the 77 (or fewer) remaining monochromatic pairs, first color the maximum number of balls a single color; this requires coloring at most 12 balls ($\binom{12}{2} = 66$) and leaves us with at most 11 monochromatic pairs to create. To create these at most 11 monochromatic pairs, we color the remaining balls in groups of 3 (3 balls give us 3 monochromatic pairs) until there are less than 3 monochromatic pairs remaining to create, at which point we require at most 4 balls (color 2 and then color 2) to create the remaining 2 monochromatic pairs. Altogether, we can create 77 monochromatic pairs using at most $12 + 3 + 3 + 3 + 4 = 25$ balls.

Therefore, as long as $\frac{1}{8}n \geq 25$, there are enough balls remaining to be colored to obtain the last 77 or fewer pairs. This is equivalent to $n \geq 200$, so we are done.

Solution 2: As in Solution 1, we want to prove that $\frac{1}{2}\binom{n}{2}$ can be written as a summation $\sum \binom{a_i}{2}$ where $\sum a_i \leq n$. We use the following lemma.

Lemma: For all positive integers N, N can be written as a summation of triangular numbers, $\sum_{i=1}^{k} \binom{a_i}{2}$, such that

$$\sum_{i=1}^{k} a_i \leq 4 + \sqrt{2N} + \sqrt[4]{128N}.$$

Proof: To prove the lemma, we write $N = \binom{a}{2} + \binom{b}{2} + c\binom{3}{2} + d\binom{2}{2}$, where a is first chosen maximally, then b is chosen maximally, then c is chosen maximally, and finally d is chosen to pick up everything else. Since a is chosen maximally, $\binom{b}{2} \leq a - 1$. Since b is chosen maximally, $3c + d \leq b - 1$. And finally $d \leq 2$. The sum of all the binomial indices is

$$\sum_{i=1}^{k} a_i = a + b + 3c + 2d \leq a + b + (b - 1) + 2 = a + 2b + 1.$$

Now $\binom{a}{2} \leq N$ implies that $\frac{1}{2}(a-1)^2 \leq N$, so $a \leq 1 + \sqrt{2N}$. Similarly, $\binom{b}{2} \leq a - 1 \leq \sqrt{2N}$ implies $b \leq 1 + \sqrt{2\sqrt{2N}}$. So the sum of all the binomial indices is

$$\sum_{i=1}^{k} a_i \leq (1 + \sqrt{2N}) + 2(1 + \sqrt{2\sqrt{2N}}) + 1 = 4 + \sqrt{2N} + \sqrt{128N}.$$

$\square$

Now that the lemma is proven, it only remains to show that for $n \geq 200$, $N = \frac{1}{2}\binom{n}{2}$ satisfies $4 + \sqrt{2N} + \sqrt[4]{128N} \leq n$. Since $N \leq \frac{1}{4}n^2$, we find

$$4 + \sqrt{2N} + \sqrt[4]{128N} - n \leq 4 + \sqrt{n^2/2} + \sqrt[4]{32n^2} - n,$$

so it suffices to show that for $n \geq 200$,

$$4 + \sqrt{n^2/2} + \sqrt[4]{32n^2} - n \leq 0. \tag{1}$$

This function is strictly decreasing, because if $t = \sqrt{n}$, it can be written as $(1/\sqrt{2} - 1)t^2 + \sqrt[4]{32}t + 4$, which is a downward opening parabola with a vertex at $t = \frac{\sqrt[4]{32}}{2-\sqrt{2}} < \frac{3}{1/2} = 6$. Hence it is decreasing for $t > 6$, or $n > 36$. So we now plug in $n = 200$, finding

$$4 + 100\sqrt{2} + 40\sqrt[4]{0.5} - 200 < 4 + 142 + 40 - 200 < 0.$$

Thus (1) is true for all $n \geq 200$, which shows that we can complete the process for all $n \geq 200$. In fact, the smallest integer value of n for which (1) is true is 92, so this actually works for $n \geq 92$.

Note: Although the problem asks to show this only for $n \geq 200$, this turns out to be possible for all $n \geq 0$ with $n \equiv 0$ or $n \equiv 1 \pmod 4$, with only three exceptions: $n = 5$, $n = 8$, and $n = 12$. In particular, as long as $n \geq 13$, it's possible to color the balls such that the probability of two balls being the same is equal to the probability of two balls being different.

12. *(2017 P6, Solved By: 0%, Average Score: 0.1, Author: Grant Molnar)*

Solution 1: We claim that $a_1 \equiv 1$, $a_2 \equiv 1$, $a_3 \equiv 2$; for $n \geq 4$ even, $a_n \equiv 1 \pmod 4$; for $n \geq 5$ odd, $a_n \equiv 0 \pmod 4$.

For the proof, begin by noting that $a_1 = 1$, $a_2 = 1$, $a_3 = 2$. Now we have two cases: $n \geq 4$ even, and $n \geq 5$ odd.

Case 1: $n \geq 5$ is odd.

The number of ways to order the deck, a_n, is equal to the number of ways to place numbers 1 through $n + 1$ around a circular table with $n + 1$ seats, such that every number is either less than its two neighbors or greater than its two neighbors, up to rotation. To see this, consider the following bijection: fix the place of $n + 1$ around the table, and then put the deck in order clockwise, starting by placing the first card in the deck to the left of $n + 1$, and ending by placing the last card in the deck to the right of $n + 1$.

Now consider any valid placement of the numbers around the table, and instead of fixing the place of $n + 1$, fix the place of 3, which is not equal to $1, 2, n$, or $n+1$. 1 and 2 cannot be adjacent around the table, or else 2 would not be higher than both neighbors; similarly n and $n + 1$ cannot be adjacent around the table, or else n would not be lower than both neighbors. Since they are not adjacent, 1 and 2 can be switched and yield a valid ordering, and n and $n + 1$ can be switched as well. Both possible switches yield 4 possibilities, meaning that every ordering around the table belongs to a set of 4 equivalent orderings when you do these two switches, and all 4 of these orderings are different since the positions of $1, 2, n$, and $n+1$ are different relative to 3. So the total number of orderings around the table is divisible by 4, i.e., $a_n \equiv 0 \pmod 4$.

Case 2: $n \geq 4$ is even.

Pair up the cards into $n/2$ *pairs:* 1 and 2, 3 and 4, 5 and 6, and so on, up to $n - 1$ and n. If the two cards in a pair, $2i - 1$ and $2i$, are not adjacent in the deck, then they are interchangeable; we can switch them and we still get a valid ordering. Additionally, switching them does not change which pairs are adjacent. Thus considering all possible ways to switch the non-adjacent pairs, we see that a single ordering is part of a set of 2^k orderings, where k is the number of non-adjacent pairs.

This is divisible by 4 unless $k \leq 1$, so it remains to count orderings of the deck where at most one pair is non-adjacent. Say that only $2i - 1$ and $2i$ are possibly not adjacent. In order for 1 and 2 to be adjacent, 2 must be higher than its neighbors, but it can only be higher than 1, so the ordering must end with $1, 2$. In order for 3 and 4 to be adjacent, in turn, 4 must be higher than its neighbors, so the ordering must end $3, 4, 1, 2$. This extends to all cards lower than $2i - 1$ and $2i$, so we see that the ordering ends with $2i - 3, 2i - 2, \ldots, 3, 4, 1, 2$. On the other hand, in order for n and $n - 1$ to be adjacent, $n - 1$ must be lower than its neighbors, and a similar reasoning shows the ordering must begin with $n - 1, n, n - 3, n - 2, \ldots, 2i + 1, 2i + 2$. The only way to then fill in the middle two cards is for $2i - 1$ and $2i$ to be also adjacent, in the middle, in that order. So the entire deck ordering must be

$$n - 1, n, n - 3, n - 2, n - 5, n - 4, \ldots, 5, 6, 3, 4, 1, 2.$$

Thus there is only 1 ordering left to count, namely, the one above. So $a_n \equiv 1 \pmod 4$.

Solution 2: Suppose that we define $a_0 = 1$. We claim that for $n \geq 0$,

$$a_n \equiv \begin{cases} 0 \pmod 4 & \text{if } n \geq 5 \text{ and } n \text{ odd}, \\ 1 \pmod 4 & \text{if } n = 1 \text{ or } n \text{ even}, \\ 2 \pmod 4 & \text{if } n = 3. \end{cases} \tag{1}$$

To prove this, we show that the following recurrence relation must hold:

$$a_{n+1} = \binom{n}{1} a_1 a_{n-1} + \binom{n}{3} a_3 a_{n-3} + \binom{n}{5} a_5 a_{n-5} + \cdots. \tag{2}$$

First we prove (2). The condition we are given in the problem can be restated as "Every card in an odd position is smaller than its neighbors." In particular, this shows that the highest card (in this case, Card $n + 1$) must appear in an even position. If Card $n + 1$ appears in position $2k$, then there are $2k - 1$ cards before it and $n - (2k - 1)$ cards after it (possibly 0 cards). We can choose which cards appear before it in $\binom{n}{2k-1}$ ways, and we can arrange these $2k - 1$ cards in a_{2k-1} ways. The remaining cards appear after Card $n + 1$, and we see that they can be arranged in $a_{n-(2k-1)}$ ways. Therefore, the number of ways to arrange $n + 1$ cards such that Card $n + 1$ appears in position $2k$ is $\binom{n}{2k-1}a_{2k-1}a_{n-(2k-1)}$. Summing this over all k, we obtain the recurrence relation (2).

Now we prove (1) by strong induction. As a base case, observe that $a_1 = 1$, $a_2 = 1$, and $a_3 = 2$. Then suppose that (1) holds for all $n \leq 2k - 1$, where $2k - 1 \geq 3$. By (2),

$$a_{2k} = \binom{2k-1}{1}a_1 a_{2k-2} + \binom{2k-1}{3}a_3 a_{2k-4} + \cdots$$

As $a_{2j-1} \equiv 0 \pmod 4$ for $2j - 1 \geq 5$, we know that the sum of all terms beyond these two terms will be $0 \pmod 4$. Therefore,

$$a_{2k} \equiv \binom{2k-1}{1}a_1 a_{2k-2} + \binom{2k-1}{3}a_3 a_{2k-4} \pmod 4$$

Also, $a_1 = 1$, $a_3 = 2$, and as $2k - 2$ and $2k - 4$ are even, we know $a_{2k-2} \equiv a_{2k-4} \equiv 1 \pmod 4$. Therefore,

$$\begin{aligned}
a_{2k} &\equiv \binom{2k-1}{1} + \binom{2k-1}{3}2 \pmod 4 \\
&= \frac{3(2k-1) + (2k-1)(2k-2)(2k-3)}{3} \\
&= \frac{(2k-1)(4k^2 - 10k + 9)}{3} \\
&\equiv \frac{(2k-1)(2k+1)}{3} \pmod 4 \\
&= \frac{4k^2 - 1}{3} \\
&\equiv \frac{-1}{3} \pmod 4 \\
&\equiv 1 \pmod 4.
\end{aligned}$$

Thus if (1) holds for all $n \leq 2k - 1$, it must hold for $n = 2k$.

Similarly, suppose that (1) holds for all $n \leq 2k$, where $2k \geq 4$. Then by (2),

$$a_{2k+1} = \binom{2k}{1}a_1 a_{2k-1} + \binom{2k}{3}a_3 a_{2k-3} + \cdots . \tag{3}$$

If $2k + 1 = 5$, then we can apply (3) to find that $a_5 = 16$. Similarly, if $2k + 1 = 7$, we find $a_7 = 272$. In both cases, $a_{2k+1} \equiv 0 \pmod 4$. Otherwise, if $2k + 1 \geq 9$, then in (3) we find that we will have at least four terms. In particular, in each product of the form $a_{2j-1}a_{2k-(2j-1)}$, at least one of $2j - 1$ and $2k - (2j-1)$ will be greater than or equal to 5 (they are odd numbers summing to $2k = (2k+1) - 1 \geq 9 - 1 = 8$). Therefore, at least one of a_{2j-1} and $a_{2k-(2j-1)}$ will be 0 (mod 4) by the inductive hypothesis in (1). Thus each of the terms $\binom{2k}{2j-1} a_{2j-1} a_{2k-(2j-1)}$ in (3) will be 0 (mod 4), hence

$$a_{2k+1} \equiv 0 \pmod 4.$$

Therefore, if (1) holds for all $n \leq 2k$ with $2k \geq 4$, it must hold for $n = 2k+1$.

Therefore, by induction, (1) must hold for all nonnegative integers n.

13. *(2013 P6, Solved By: 0%, Average Score: 0.1, Author: Hiram Golze)*

First, we claim that the number of ways to tile the $1 \times n$ hexagonal strip below is F_n, the nth Fibonacci number, where $F_0 = F_1 = 1$ and $F_{n+1} = F_n + F_{n-1}$.

n tiles

This is obviously true for $n = 1$ and $n = 2$ because we have 1 way to tile a 1×1 hexagonal strip and we have two ways to tile a 2×1 hexagonal strip. Now, suppose the statement is true for some k and some $k+1$. Then a $(k+2) \times 1$ hexagonal strip can end in either consist of a $(k+1) \times 1$ strip followed by a 1×1 strip or a $k \times 1$ strip followed by a 2×1 strip. Thus the number of ways to tile a $(k+2) \times 1$ strip is $F_k + F_{k+1} = F_{k+2}$, so in fact the statement is true for all n, as desired.

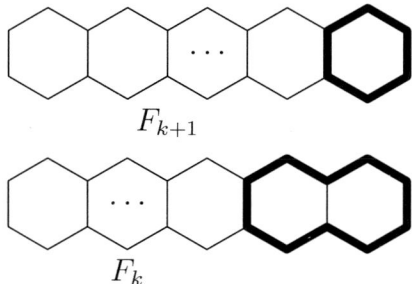

Now we claim that the number of ways to tile the hexagonal border of a triangle where each side is of length n is F_n^3. To demonstrate this, it suffices to demonstrate a one-to-one correspondence between tilings of the hexagonal border and ordered 3-tuples of tilings of a $n \times 1$ hexagonal strip, or in other words, the number of ways to tile three $n \times 1$ hexagonal strips, one colored red, one colored blue, and one colored green.

For a given tiling of the hexagonal border of a triangle, we attempt to map the tiling of the bottom $n \times 1$ strip to a red tiling, the tiling of the left $n \times 1$ strip to a blue tiling, and the tiling of the right $n \times 1$ strip to a green tiling by "pulling" the tiling off. For example, we can map the following tiling of a 4-hexagonal grid as shown below to an RGB $n \times 1$ tiling.

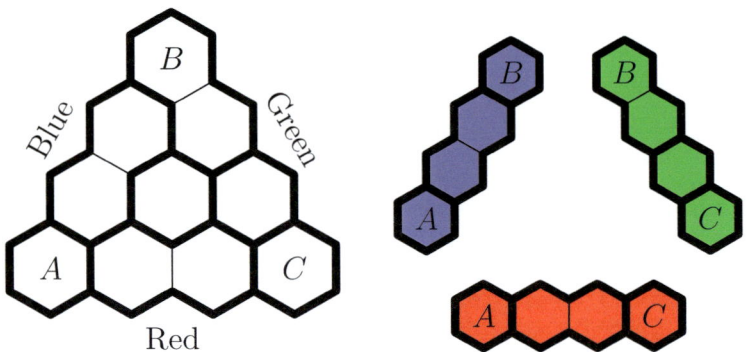

This can be done in the natural way if the bottom, right, and left sides consist of valid $n \times 1$ tilings. But this doesn't always happen. The only case where it can fail, however, is if we have a corner tiled as shown below. In this case, we map it to the shown red and green tilings.

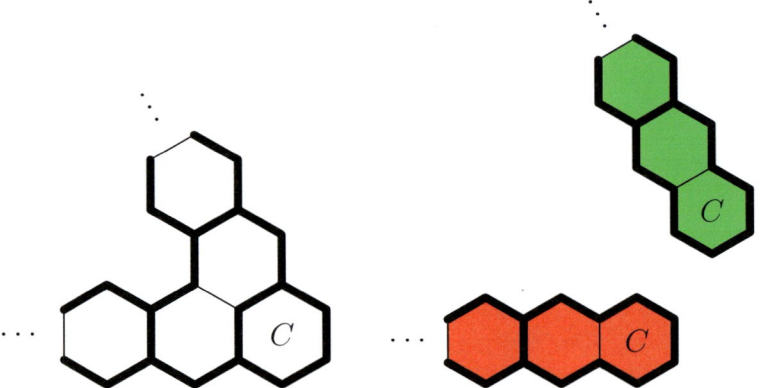

It is easy to see that this will map every tiling of the n-hexagonal grid $(n \geq 3)$ to a distinct RGB n-tiling. Furthermore, it is also easy to see that every RGB n-tiling will come from a valid tiling of the n-hexagonal grid. Therefore, this gives us a one to one correspondence between RGB n-tilings and tilings of the n-hexagonal grid. From our work at the beginning, the number of red, green, and blue n-tilings is F_n, so the number of RGB n-tilings is just F_n^3, which is therefore the number of tilings of the n hexagonal grid.

Note 1: For example, the first few terms are (for a diagram of side length $n \geq 3$)

$$27, 125, 512, 2197, 9261, 39304, 166375, \ldots$$

and is given by sequence A056570 in OEIS [9].

Note 2: We originally considered a version of this problem on a *filled* triangular grid of hexagons, rather than just the border. This version of the problem is much harder, and does not seem to admit a simple closed-form solution. The first few terms are (for a grid of side length $n \geq 3$)

$$27, 425, 14278, 1054396, 169858667, 59811185171, \ldots$$

and is given by sequence A269869 in OEIS [10].

3.3 GAMES

1. *(2016 Flyer, Author: Samuel Dittmer)*

   Malcolm wins.

   We first claim that the game is equivalent to the game with the following list of numbers available instead:

   $$2, 2, 2, \quad 3, 3, \quad 5, \quad 0, 0, 0.$$

   To see this, note that $2, 4, 8$ and $3, 9$ are equivalent, as they are divisible by the same set of primes. Additionally, note that all of the numbers available are divisible by at least one of $2, 3,$ or 5. Therefore, in order for three numbers to be pairwise relatively prime, they must each be divisible by *only* one of $2, 3,$ or 5. Therefore, $6, 10,$ and 15 simply act as blocking numbers and prevent Malcolm from winning in whatever row, column, or diagonal they are placed in. We can replace them with 0, since 0 is not relatively prime with any other integer other than 1.

   With this in mind, we describe a winning strategy for Malcolm. He begins by placing 2 in the center. If Ozymandias responds with 3 or 5, Malcolm can win by placing 5 or 3 (respectively) opposite Ozymandias's move. Also, if Ozymandias responds with 2, Malcolm can always create a "fork" with 5 and win in two more moves:

   (Since there are two 3s available, wherever Ozymandias plays next, Malcolm can play 3 in the bottom left, bottom right, or middle right to win.)

So Ozymandias's only hope is to respond with a 0. There are two cases: Ozymandias plays in a corner, or on an edge. In either case, we claim that Malcolm can win by placing a 2 opposite from the 0.

Corner: After Ozymandias plays the 0 in a corner and Malcolm plays the opposite 2, we arrive at the following board:

Consider Ozymandias's next move. A 3 or 5 loses as Malcolm can respond with a 5 or 3 to win. For any other number, Malcolm is threatening one of two possible forks, depending on where Ozymandias played (marked ? below):

Regardless, following the fork, Malcolm will win on the next move by placing a 3. So Malcolm wins in this case.

Edge: On the other hand, if Ozymandias plays the 0 on an edge, after Malcolm responds we arrive at the following board:

As before, 3 and 5 will lose to a 5 or 3, respectively, so Ozymandias must play a 0 or 2. Additionally, a 0 or 2 vertically or diagonally adjacent to the currently placed 0 also loses, due to another fork threat:

So Ozymandias must place in the top-right or bottom-right corner; without loss of generality, take the top-right. If Ozymandius plays a 2, Malcolm responds with yet another fork:

The only remaining—and perhaps most interesting—case is where Ozymandias responds with a 0 in the corner. Here, Malcolm does not have any fork threats. However, he can still win, we claim, by placing a 3 in the top left:

$$\begin{array}{c|c|c} 3 & & 0 \\ \hline 0 & 2 & 2 \\ \hline & & \end{array}$$

The idea is that Malcolm is threatening to win with a 5 on the bottom right, but *also* threatening to win in a new indirect way: by placing a 0 in the top center. This is because if a 0 is played in the top center, the three remaining numbers are $\{2,3,5\}$ to be placed across the bottom row, so Malcolm will always win.

How can Ozymandius prevent both threats? There are only two possibilities: by placing a 5 in the top center, or by placing a 0 in the bottom right. But if Ozymandias plays the 5 in the top center, Ozymandias wins with a 3 in the bottom center:

$$\begin{array}{c|c|c} 3 & 5 & 0 \\ \hline 0 & 2 & 2 \\ \hline & 3 & \end{array}$$

So the only remaining option is for Ozymandias to block the diagonal with a 0, at which point we get this picture:

$$\begin{array}{c|c|c} 3 & & 0 \\ \hline 0 & 2 & 2 \\ \hline & & 0 \end{array}$$

To complete this last case, Malcolm places the winning move of a 2 in the bottom left square, ensuring that the only remaining numbers are $\{3,5\}$ and guaranteeing a win along the center column.

$$\begin{array}{c|c|c} 3 & & 0 \\ \hline 0 & 2 & 2 \\ \hline 2 & & 0 \end{array}$$

Therefore, Malcolm wins in any continuation after placing 2 in the center square. This completes the winning strategy.

2. *(2021 Flyer, Author: Grant Molnar)*

Bonnie wins. Here is a way to see the game visually, where m covers n if and only if there is a path of arrows from m to n:

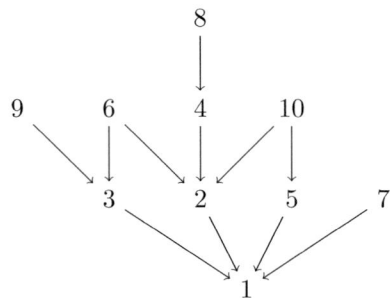

We can show that Bonnie wins by a strategy-stealing argument. Notice that 1 is covered by every other integer, since every integer is a multiple of 1. Hence, 1 can only be named on the very first turn, since it will always be covered on future turns. Supposing it were true that Clyde had a winning strategy, then Bonnie could first name 1, then follow Clyde's supposed strategy on all of Clyde's future moves, and win. Thus, Clyde cannot have a winning strategy. But this doesn't show *how* Bonnie can win, only that she must have some way to do so.

In fact, Bonnie can win by naming 6. After naming 6, $1, 2, 3$, and 6 get covered, so the remaining moves in the game are: $8 \to 4$, $10 \to 5$, 9, and 7. From this point, the game is symmetric: Bonnie can respond to Clyde's moves on $8 \to 4$ on $10 \to 5$ and vice versa, and respond to 9 with 5 and vice versa. So Clyde will be the first to be unable to win, and Bonnie wins.

Note: Choosing 4 at the beginning also wins. All other moves lose.

3. *(2014 P1, Solved By: 46%, Average Score: 4.4, Author: Hiram Golze)*

First, note that both Todd and Allison place exactly one blue stone in each of their turns. Also, it will take Todd at least eight moves to get from the center to a corner. Therefore, as Todd would go both first and last, it would take at least $8 + 7 = 15$ turns of Todd and Allison combined before Todd could move to a corner square. Therefore, at least 15 blue stones are placed on the grid.

We claim that Allison can always force Todd to move to a corner square in his eighth turn. Without loss of generality, we may assume that Todd's first turn is to move north. Allison decides that she wants to force Todd to the northeast corner, namely corner D. In order to do this, she plays immediately to the West of Todd, blocking any westward motion. So after Todd and Allison's first turns, we may assume that the board looks like the following diagram.

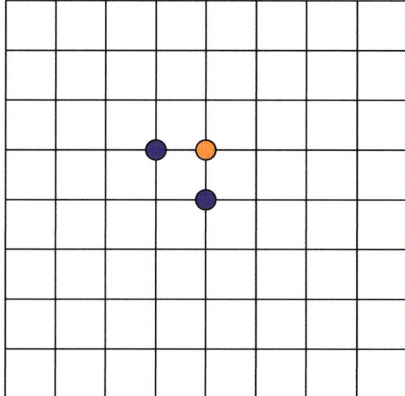

Note that Todd can only move north or west. After this point, Allison executes the following strategy based on Todd's turn.

(i) If Todd moves north, then Allison plays directly to his west.

(ii) If Todd moves east, then Allison plays directly to his south.

In order for this strategy to make sense, we must show that after Allison moves, then Todd can only move to the north or to the west. Let's look at a very general picture. If Todd moves north, then he leaves a blue stone behind him, and Allison plays to his west. This leaves the following arrangement of the board.

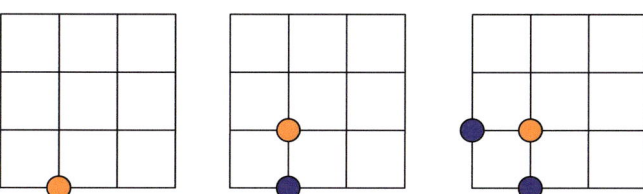

Therefore, in this case, Todd can only move north or east.

In the case that Todd moves to the east, then Allison moves directly to his south, so we obtain the following picture.

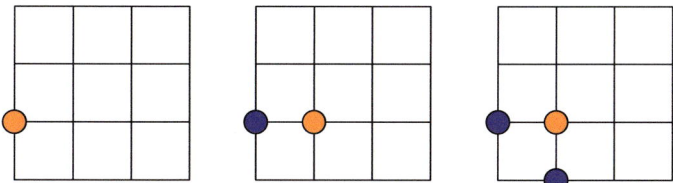

Once again, it is clear that Todd can only move to the north or to the east.

Therefore, after Allison moves according to her strategy, Todd can only move in the north or east direction. But he can only make 4 north moves before reaching the edge of the board, and 4 east moves before reaching the edge of the board. Therefore, according to the described strategy of Allison, she can force Todd to reach a corner in 8 steps. This is also the minimal number of steps that it would take Todd to reach a corner, so we conclude that this is Allison's optimal strategy. Therefore, as stated earlier, we conclude that at the end of the game, there will be exactly 15 blue stones on the board.

4. *(2016 P1, Solved By: 43%, Average Score: 3.9, Author: Grant Molnar)*

   Ada wins. Here is one possible strategy.

   Turn 1: Ada begins by putting the pebble at A.

   Turn 2: Ada removes stick AB, and places the pebble at B.

   Turn 3: The only remaining stick for Otto to choose is stick BC, so Otto removes this stick, placing the pebble at C.

   Turn 4: Ada removes stick CD, placing the pebble at D.

   Turn 5: Otto is again forced to choose the only remaining adjacent stick, DE. So Otto removes this stick, and places the pebble at E.

   Turn 6: Ada removes stick EF, placing the pebble at F.

   Turn 7: Otto is forced to remove stick FA, placing the pebble at A.

   Turn 8: Ada removes stick AC, placing the pebble at C.

   Turn 9: Otto is forced to remove stick CE, placing the pebble at E.

   Turn 10: Ada removes stick EA, leaving no remaining sticks.

   At this point, there are no valid moves, so Otto loses. Since every one of Otto's moves above is forced, this is a winning strategy.

5. *(2018 P3, Solved By: 36%, Average Score: 3, Author: Grant Molnar)*

   We claim that Kim has a winning strategy. We demonstrate one possible strategy below. Kim's goal in this strategy is to guarantee the ability to place 4 tiles if necessary, one in each column, while at the same time limiting Li to placing at most 3 tiles. This allows Kim to make the final move.

   Kim begins by placing a tile in the middle of the second column. Then, no matter where Li places next, Kim places a tile in the third column, either in the top two squares or the bottom two squares—this is possible since Li's move can't have blocked both. The diagram below shows the position of Kim's first two moves as blue tiles.

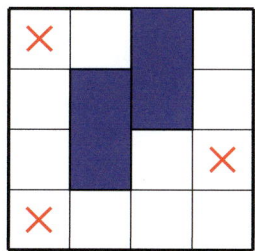

Notice that at this point, two adjacent cells in the first column and two adjacent cells in the fourth column are free, such that Li can never place a rectangle in them; so Kim is already guaranteed the ability to make two more moves, or four moves total. However, Kim still has to make sure that Li can only place 3 tiles total. For this, consider three squares, marked as red Xs above: the top left corner, the bottom left corner, and the square in the middle two squares of the fourth column which is not adjacent to Kim's previously placed tile. Li has only placed two moves so far, so Li could not have covered all three Xs. For the third move, Kim places either in the first column or in the fourth column to block one of the Xs.

After Kim's third move, there are still the four squares (in the first and fourth columns) that Li cannot place a rectangle on as noted above. But there is now one additional square (one of the Xs) that is blocked. Therefore, including Kim's initial moves, there are $4 + 4 + 1 = 9$ squares that Li cannot place a rectangle on in the course of this game, or rather $16 - 9 = 7$ squares that Li might be able to place a rectangle on. Thus Li can only place at most 3 tiles total. But Kim still has a fourth move available (either in the first or fourth column), so Kim wins.

6. *(2019 P4, Solved By: 30%, Average Score: 2.2, Author: Caleb Stanford)*

We claim that the leader has a winning strategy. That is, it is possible for the leader to give instructions to the players such that the leader always wins, no matter which player is the traitor and no matter how the traitor plays.

Here is one example of instructions that work. Number the players from 1 to 10, and assume that the leader passes first to player 1. For $i < 10$, let player i pass to player $i + 1$ on the first pass, and to player 10 on all passes after that. Then let player 10 pass first to player 9, then to player 8, then to player 7, and so on. These instructions are illustrated below.

Player 1 Instructions:	Player 2 Instructions:	Player 3 Instructions:
1. Pass to Player 2	1. Pass to Player 3	1. Pass to Player 4
2. Pass to Player 10	2. Pass to Player 10	2. Pass to Player 10
3. Pass to Player 10	3. Pass to Player 10	3. Pass to Player 10
⋮	⋮	⋮

Player 4 Instructions:	Player 5 Instructions:	Player 6 Instructions:
1. Pass to Player 5	1. Pass to Player 6	1. Pass to Player 7
2. Pass to Player 10	2. Pass to Player 10	2. Pass to Player 10
3. Pass to Player 10	3. Pass to Player 10	3. Pass to Player 10
⋮	⋮	⋮

Player 7 Instructions:	Player 8 Instructions:	Player 9 Instructions:
1. Pass to Player 8	1. Pass to Player 9	1. Pass to Player 10
2. Pass to Player 10	2. Pass to Player 10	2. Pass to Player 10
3. Pass to Player 10	3. Pass to Player 10	3. Pass to Player 10
⋮	⋮	⋮

Player 10 Instructions:
1. Pass to Player 9
2. Pass to Player 8
3. Pass to Player 7
⋮

We claim that with these instructions, the leader wins. If player 10 is the traitor, the leader wins as soon as the traitor gets the ball for the first time. If player ℓ is the traitor, for $\ell < 10$, then players 1 through $\ell - 1$ will have already gotten the ball when the traitor gets it. No matter who the traitor throws the ball to next, the ball will get to player 10 before it gets back to player ℓ. Player 10 then throws to players $9, 8, \ldots, \ell + 1$, in order, who each throw the ball back to player 10. Therefore, all of the remaining players will receive the ball before player ℓ (the traitor) gets the ball again.

7. *(2013 P2, Solved By: 13%, Average Score: 1.8, Author: Hiram Golze)*

We claim that Alice has a winning strategy if and only if $1 \leq A < 2$ or $2 < A \leq 4$. To show this, consider cases on A:

- If $1 \leq A < 2$, then Alice wins because she can cut the paper in half. Carl receives a sheet of paper of area less than 1 and loses.

- If $A = 2$, then Alice can only cut off any area up to half, so the remaining area will be greater than or equal to 1, but less than 2. Carl may then cut off half, giving Alice a sheet of area less than 1. Therefore, in this case, Carl wins.

- If $2 < A \leq 4$, then Alice can cut the paper into sheets of size 2 and $A - 2$. Carl will receive the sheet of size 2, and loses by the logic in the $A = 2$ case. Therefore, Alice has a winning strategy.

- Finally, we claim that for all other areas above 4, the optimal strategy leads the game to go on infinitely (that is, neither player has a winning strategy). First, consider what happens when a player receives a sheet of paper with area $4 < A \leq 8$. Then when they make their cut, the paper that they pass will have area greater than 2 and less than 8. But if they give the other player a piece of paper of area $2 < A \leq 4$, then as noted in the previous case, that player will have a winning strategy. Therefore, no matter the starting area, no player will ever make a cut that leaves the other player with an area less than 4, so the game will go on indefinitely.

Note: There was a minor error in the originally published version of this problem and its solution. To correct the error, we added the condition $A \geq 1$.

8. *(2017 P5, Solved By: 11%, Average Score: 1.7, Author: Caleb Stanford)*

   We can defeat the Great Pumpkin. Proceed in two phases.

   - For phase 1: begin by pouring an equal amount, $\frac{1}{8}$, into every bucket. When the Great Pumpkin pours out two, refill those (taking $\frac{1}{4}$), and then distribute the remaining water ($\frac{3}{4}$) equally between all 8 buckets. On every turn, continue refilling the two buckets that the Great Pumpkin emptied, and distributing equally the remaining water.

     Suppose that after our turn all buckets are at r and $r < \frac{1}{2}$. Then the Great Pumpkin empties out two buckets with r liters on its turn. On our next turn we take $2r$ to refill the two emptied buckets, and then distribute $1 - 2r$ equally, bringing all buckets to

     $$r + \frac{1 - 2r}{8} = \frac{3}{4}(r) + \frac{1}{4}\left(\frac{1}{2}\right).$$

     This is a weighted mean, and it expresses the number one-quarter of the way from r to $\frac{1}{2}$. Thus by repeating this strategy, the gap between r and $\frac{1}{2}$ will become smaller by a factor of $\frac{3}{4}$ each time. Since $\left(\frac{3}{4}\right)^n$ approaches 0 as n grows to infinity, we can get as close as we want to $\frac{1}{2}$ in all buckets.

     Stay in Phase 1 until the amount of the water in each bucket is greater than $\frac{7}{15}$—this is still less than $\frac{1}{2}$, but it will be close enough to win. Then move on to Phase 2.

   - For phase 2: on your turn, six buckets have x liters of water in them, where $x > \frac{7}{15}$. Pick five of them, and fill in $\frac{1}{5}$ of a liter in each.

     On your next turn, at least 3 buckets are left with $x + \frac{1}{5}$. Fill in $\frac{1}{3}$ of a liter in each.

Finally, on your following turn, at least 1 bucket is left with $x + \frac{1}{5} + \frac{1}{3}$. Fill the entire liter into this bucket.

The resulting amount of water in the bucket is

$$x + \frac{1}{5} + \frac{1}{3} + 1 = x + \frac{23}{15} > \frac{7}{15} + \frac{23}{15} = 2.$$

Since the bucket is only able to hold 2 liters, we have caused it to overflow and we win.

Note: In December 2020, this problem was included in Peter Winkler's latest book: *Mathematical Puzzles* [21]. It was inspired by a problem proposed by Gerhard Woeginger for the 2009 International Math Olympiad shortlist, featuring Cinderella and her wicked stepmother with five buckets in a pentagon shape (see [7, 3]).

9. *(2021 P5, Solved By: 14%, Average Score: 1.6, Author: Caleb Stanford)*

We claim that each player has a strategy that prevents them from losing, hence if both players play optimally, then the game will go on infinitely. The strategy to prevent yourself from losing is simple: if you have 0 or 1 stones on your turn, then gain a stone (this is forced); otherwise if you have $n \geq 2$ stones, then always give away as many stones as possible ($\lfloor \frac{n}{2} \rfloor$, where $\lfloor \cdot \rfloor$ is the floor function) to the opponent.

We argue that this strategy works for Gog; the argument that it works for Magog is identical. Suppose that it is our turn and we have not lost yet; then we have at most 19 stones, and after giving as many as possible away (or gaining one in the case of 0 or 1), we will have at most 10 stones. Then either the opponent loses immediately, or on their turn they can give us at most 9 stones. Since that leaves us with at most $10 + 9 = 19$ stones again on our turn, we do not lose and by induction, we can never lose.

10. *(2022 P4, Solved By: 6%, Average Score: 1.5, Author: Annie Yun)*

We claim that Alpha has a winning strategy. To prove this, define a new game where the position denoted (m, n) corresponds to an $(m - 1) \times (n - 1)$ grid of squares in the original game, so that the starting position is $(11, 101)$. That is, we add one to the two coordinates' names, but the actual moves are the same. Then in the new game, a move $(m, n) \to (m', n)$ is legal if and only if $\frac{m}{2} < m' < m$, and similarly for a move $(m, n) \to (m, n')$. This is because folding the paper either reduces the width by at most half, or the length by at most half, which corresponds to reducing the coordinates in the new game by strictly less than half.

Now we claim that in the new game, the losing positions (where the second player wins) are exactly the positions of the form $(a, 2^k a)$ or

$(2^k a, a)$, for some $a \geq 2$ and $k \geq 0$—that is, positions where the ratio is a power of two. For example, $(10, 20), (10, 40)$, and $(10, 80)$ are losing, while $(10, 30)$ is winning.

First, we describe a winning strategy for player 2 in such situations. Given a position $(a, 2^k a)$, there are two cases for the first player's move: either she moves to $(a', 2^k a)$ (where $a/2 < a' < a$), or she moves to (a, j) (where $2^{k-1} a < j < 2^k a$). In the first case, we respond by moving to $(a', 2^k a')$, which is legal because $2^k a' > 2^k (a/2) = (2^k a)/2$, and it preserves the form where one coordinate is a power of two times the other. In the second case, we respond by moving to $(a, 2^{k-1} a)$, which is legal because j is an integer, so $j < 2^k a - 1$, so $j/2 < 2^{k-1} a - 1/2$, so $2^{k-1} a > j/2$. Ultimately, such a strategy is possible as long as $a \geq 2$ and $k \geq 1$, and it will continue until the first player is unable to make a move (position $(2, 2)$), at which point the second player wins.

Now we argue that there is a winning strategy for player 1 in all other positions. Let the position be (a, b), where without loss of generality $a < b$ (note that $a = b$ is a winning position for player 2). Then let $k \geq 0$ be the unique integer such that $2^k < \frac{b}{a} < 2^{k+1}$—we know that such a k exists because the position is not of the form $a, 2^k a$. Then, player 1's strategy is to move to $(a, 2^k a)$. From here, player 1 can win with player 2's strategy from before.

To complete the proof, since the starting position $(11, 101)$ has a ratio $\frac{101}{11}$, which is not a power of two (not even an integer), it follows that the position is winning, and Alpha has a winning strategy. In particular, Alpha can win by going to the losing position $(11, 88)$. In the original terms, Alpha can win by folding the 10×100 grid into a 10×87 grid.

Note: Concretely, for $m \leq 11$, $n \leq 101$, and $m \leq n$, the losing positions are:

$$(1, 1), (1, 2), (1, 4), (1, 8), (1, 16), (1, 32), (1, 64)$$
$$(2, 2), (2, 4), (2, 8), (2, 16), (2, 32), (2, 64)$$
$$(3, 3), (3, 6), (3, 12), (3, 24), (3, 48), (3, 96)$$
$$(4, 4), (4, 8), (4, 16), (4, 32), (4, 64)$$
$$(5, 5), (5, 10), (5, 20), (5, 40), (5, 80)$$
$$(6, 6), (6, 12), (6, 24), (6, 48), (6, 96)$$
$$(7, 7), (7, 14), (7, 28), (7, 56)$$
$$(8, 8), (8, 16), (8, 32), (8, 64)$$
$$(9, 9), (9, 18), (9, 36), (9, 72)$$
$$(10, 10), (10, 20), (10, 40), (10, 80)$$
$$(11, 11), (11, 22), (11, 44), (11, 88).$$

11. *(2020 P6, Solved By: 0%, Average Score: 0.7, Author: Caleb Stanford)*

The answer is that player 6 wins. This is trickier to prove than to state, and many intuitive justifications are not fully rigorous, so we have to be careful.

Solution 1: First, we show that one of players $1, 2, 3, 4, 5,$ or 6 must win. Second, we show that one of players $6, 7, 8, 9,$ or 10 must win. Logically, it follows that 6 must win.

- For the first part: we claim that, for $i = 1, 2, 3, 4, 5, 6$, if players 1 through $i - 1$ all vote for 6, then one of players i through 6 wins. The proof is by backward induction. First, if players 1 through 5 all vote for 6, then player 6 can vote for themselves and win, and this is their top choice so they will do so. Next, if players 1 through 4 all vote for 6, then player 5 can either vote for 6 (making 6 win), or they can do better—and the only thing better for 5 is that 5 wins. Similarly, if players 1 through 3 vote for 6, then player 4 either makes 5 or 6 win by voting 6, or can do better, making 4 win. Continuing backward until $i = 1$, we get that one of players $1, 2, 3, 4, 5,$ or 6 must win.

- For the second part: we claim that, for $i = 6, 7, 8, 9, 10$, regardless of the first 5 votes, if players 6 through $i - 1$ vote for 10, then one of players i through 10 wins. The proof is by backward induction, similar to before. For $i = 10$, we know that if players 6 through 9 vote for 10, then player 10 can vote for themselves and win. For $i = 9$, player 9 can either vote for 10, in which case player 10 wins, or do better, so either 9 or 10 must win. After continuing backward until $i = 6$, we get that regardless of the votes of players 1 through 5, one of players 6 through 10 must win.

Solution 2: In this solution, we argue that the game is equivalent to a variant game where, on your turn instead of voting, you may "pass" your vote to the next player. Then we argue that it is rational for players 1 through 5 to pass their votes to player 6, and player 6 will then win.

First, we define the variant game. Each player starts with one vote. On your turn, you can either use all your votes, or you can "pass" your votes to the next player, who will then have your votes, plus one more. That player may then either use all their votes (in any distribution—for example you can vote for two different players), or pass all their votes to the next player, and so on.

We claim that the outcome of the variant game is exactly the same as the outcome in the original game. To see this, first imagine you are player 9 and you have some number of votes. Then instead of passing your votes on to the next player, since you know that player 10 is perfectly rational, you can just predict what player 10 would have voted, and vote that way yourself. So passing your votes to the next player never gives you any advantage, and we can just as well assume that player 9 does not pass on their votes. Similarly, going backward, player 8 could just vote themselves what they know player 9 would do, instead of passing on their votes. So we can just as well assume that player 8 does not pass on their votes. Continuing in this way, we see that we can just as well assume that no player passes on their votes, and everyone is still behaving rationally in the variant game. That means that the variant game is equivalent to the original where passing was not allowed.

Now that we have defined the variant game, we argue that on any player's turn, *either that player can ensure that they win with a certain vote, or we can assume without loss of generality that they pass on their vote.* The idea is that, if you are player i you have various choices for what you can vote. Either one of these choices leads to player i winning, or else all of the choices lead to some player $j \neq i$ winning. In the latter case, *player $i + 1$'s preferences are exactly the same as player i's.* Therefore, player i is happy to pass on their vote to player $i + 1$, as they know that player $i + 1$ will make the same choice that they want.

At this point, we are assuming without loss of generality that each player's strategy is to pass on their vote, unless they have some way to win. But only one player can win. So in the actual outcome of the game, suppose player i wins. First we show $i \leq 5$ is impossible. In this case, players 1 through $i - 1$ pass on their vote (because they can't win), and player i gets i votes. After this, players $i + 1$ through 10 pass on their vote (because they can't win either), so player 10 gets $10 - i$ votes. But then player 10 can vote for themselves and win, which is a contradiction if $i \leq 5$.

Therefore, players 1 through 5 can't win, so all five of them pass on their votes. At this point, player 6 has 6 votes. So they can force themselves to win by voting for all 6 votes on themselves. This proves that player 6 wins in the variant game with our constraint on strategies; and by the logic above, that means that player 6 wins in the original game as well.

Note: This problem has an interesting alternate version: what if each players' preferences are in *counterclockwise* order from themselves, rather than clockwise (but they still take turns in clockwise order from 1 to 10)? This version of the problem appears to be significantly more challenging. For example, Solution 2 no longer applies, since player $(i + 1)$'s voting preferences do not align with i's, even under the assumption that i does not win. The authors are not currently aware of a rigorous proof for this version of the problem [18].

12. *(2013 P5, Solved By: 3%, Average Score: 0.6, Author: Samuel Dittmer)*

Malone has the winning strategy.

Solution 1: One possible winning strategy is as follows. Malone first picks $a = 1$. If Cooper picks $b = b_0$, then Malone picks $c = \frac{1}{2}(b_0^2 + 1)$. If Cooper picks $c = c_0$, then Malone picks $b = 1$.

Now we prove why this works. Suppose we have a finished game, with polynomial $x^3 + ax^2 + bx + c$. Then this polynomial can be factored as $x^3 + ax^2 + bx + c = (x - r_1)(x - r_2)(x - r_3)$. We know from Vieta's formulas (or simply expanding the factorization) that

$$-a = r_1 + r_2 + r_3$$
$$b = r_1 r_2 + r_2 r_3 + r_3 r_1$$
$$-c = r_1 r_2 r_3.$$

Using these equations, we get that $a^2 - 2b = r_1^2 + r_2^2 + r_3^2$ and $b^2 - 2ac = r_1^2 r_2^2 + r_2^2 r_3^2 + r_3^2 r_1^2$. If all of the roots are real, then both of these quantities must be greater than or equal to 0 because they are the sums of squares. Therefore, if one of these two quantities is negative, then the polynomial must have a nonreal root. Now if we implement our strategy in the first case, the two quantities are $1 - 2b_0$ and $b_0^2 - (b_0^2 + 1) = -1$, respectively. As the second quantity is negative, the cubic must have a nonreal root. In the second case, the two quantities are $1^2 - 2(1) = -1$ and $1^2 - 2c_0$, respectively. As the first quantity is negative, the cubic must have a nonreal root. In either case, the cubic has a nonreal root, so Malone wins.

Solution 2: To prove that Malone has the winning strategy, we claim that Malone can always force the cubic to factor as $(x^2 + 1)(x + s)$, where s is a real number. Clearly, this would have a nonreal root, i. This factorization expands to $x^3 + sx^2 + x + s$. Based on the expansion, we see that Malone's first move should be to pick $b = 1$. If Cooper chooses $a = a_0$, then Malone chooses $c = a_0$, yielding the cubic $x^3 + a_0 x^2 + x + a_0$. If Cooper chooses $c = c_0$, then Malone chooses $a = c_0$, yielding the cubic $x^3 + c_0 x^2 + x + c_0$. It is clear that both cubic polynomials are of the desired form, hence each must have a nonreal root. Therefore, Malone has the winning strategy.

Note: Solution 2 is due to Benjamin Lovelady.

3.4 ALGEBRA

1. *(2014 Flyer, Author: Caleb Stanford)*

We claim the maximum possible value is 1.5.

First, we show that this is possible. The following matrix has all nonzero entries 1.5 or greater. Since every row sums to 5 and every column sums to 4, the average of all rows and columns is 1, so it is balanced.

$$\begin{bmatrix} 2.5 & 2.5 & 0 & 0 & 0 \\ 1.5 & 1.5 & 2 & 0 & 0 \\ 0 & 0 & 2 & 1.5 & 1.5 \\ 0 & 0 & 0 & 2.5 & 2.5 \end{bmatrix}$$

Now we argue that it is not possible to do better. Consider any balanced 4×5 matrix; we claim it has a nonzero entry that is 1.5 or less. Consider any row in the matrix; it can't have only one nonzero entry, or that entry would be 5, violating the constraint that columns sum to 4. So it has two or more nonzero entries. Now there are two cases:

- If all rows have three or more nonzero entries, then there are at least 12 nonzero entries total, so by the pigeonhole principle, there must be a column with at least three nonzero entries. But then these three entries sum to 4, so one of them is at most $\frac{4}{3} = 1.33\ldots$, which is below 1.5.

- Otherwise, there must be a row with exactly two nonzero entries. Then one of these two entries is at least $\frac{5}{2} = 2.5$. If this entry is equal to 4, then the other nonzero entry in the same row is 1, which is less than 1.5. On the other hand, if it is less than 4, then the other nonzero entries in the same column must add to at most $4 - 2.5 = 1.5$, so one of them is at most 1.5.

In any case, any balanced 4×5 matrix must have a nonzero entry that is at most 1.5. Therefore, the minimum nonzero element is at most 1.5, so the maximum possible value of the minimum nonzero element is exactly 1.5.

Note: This problem is based on the *muffin problem* [5], also known as the *cupcake puzzle* [1]. A typical version of the puzzle asks for the best way to divide 3 muffins equally among 5 students so as to maximize the size of the smallest piece given to any student. For more on this topic, see the recent book [5].

2. *(2020 Flyer, Author: Grant Molnar)*

   We claim that for all $n \geq 1$,

   $$A_n = \frac{(n!)^2}{2^{n-1}}.$$

   This can be proven directly by induction, but let us instead show how to arrive at this formula. First, rearrange the recurrence relation for A_n as

   $$A_n - \frac{n^2}{2} A_{n-1} = \frac{-2}{(n-1)^2} \left[A_{n-1} - \frac{(n-1)^2}{2} A_{n-2} \right].$$

It follows that if we let $B_n = A_n - \frac{n^2}{2}A_{n-1}$, then

$$B_n = \frac{-2}{(n-1)^2}B_{n-1},$$

for all $n \geq 3$. In addition, $B_2 = 2 - \frac{2^2}{2} \cdot 1 = 0$. Since 0 multiplied by anything is 0, the above recurrence relation implies that $B_n = 0$ for all $n \geq 2$.

Returning to A_n, the equation $B_n = 0$ gives us, for all $n \geq 2$:

$$A_n = \frac{n^2}{2}A_{n-1}$$

which we can expand out $(n-1)$ times to get

$$A_n = \left(\frac{n^2}{2}\right)\left(\frac{(n-1)^2}{2}\right)\left(\frac{(n-2)^2}{2}\right)\cdots\left(\frac{3^2}{2}\right)\left(\frac{2^2}{2}\right)A_1.$$

Finally, substituting $A_1 = 1$ and collapsing the products on the top and bottom of the fraction, we arrive at our formula:

$$A_n = \frac{n^2(n-1)^2(n-2)^2\cdots 2^2\cdot 1^2}{2^{n-1}} = \frac{(n!)^2}{2^{n-1}}.$$

3. *(2022 Flyer, Author: Caleb Stanford and Grant Molnar)*

Yes, $a, b,$ and c must be integers. Let $\mathbb{Z}$ denote the set of integers. Plugging in $n = 0, 1,$ and 2, we find:

$$a + c \in \mathbb{Z}$$
$$2a + b + c \in \mathbb{Z}$$
$$4a + 2b + c \in \mathbb{Z}.$$

Since the difference of two integers is an integer, subtracting two times the second expression from the last expression, we get

$$(4a + 2b + 2c) - (4a + 2b + c) = c \in \mathbb{Z}.$$

Similarly, we subtract c from $a + c \in \mathbb{Z}$ to find $a \in \mathbb{Z}$. Lastly, we subtract $2a + c$ from $2a + b + c$ to find $b \in \mathbb{Z}$.

4. *(2018 P2, Solved By: 64%, Average Score: 4.9, Author: Hiram Golze)*

The minimum possible value of $a + b + c$ given the constraints is 3.

Solution 1: We can use Vieta's formulas to find the coefficients of P and Q. Equating them, we get:

$$a + b + c = ab + bc + ca$$
$$ab + bc + ca = abc(a + b + c)$$
$$abc = (abc)^2$$

The last equation implies $abc = 0$ or 1. Since a, b, c are positive, $abc = 1$. By the AM-GM inequality,

$$\frac{a+b+c}{3} \geq \sqrt[3]{abc} = 1,$$

so $a + b + c \geq 3$. Conversely, 3 is achieved when $a = b = c = 1$, and in this case $P(x) = Q(x)$ so all constraints are satisfied.

Solution 2: Since polynomial factorization is unique, we know that the three roots a, b, c must be equal to ab, bc, ca (but possibly not in the same order). We first argue that one of the roots is 1. If $a = ab$, then since $a \neq 0$ we get $b = 1$; similarly we are done if $x = xy$ for any x, y. The only remaining case is if $a = bc$, $b = ca$, and $c = ab$. In this case, multiply the three equations together to get $abc = (abc)^2$, so $abc = 1$, and since $a = bc$, $a^2 = 1$, so $a = 1$.

Therefore, one of the roots is 1, and without loss of generality $a = 1$. Then $1, b, c$ must be equal to b, c, bc, which implies $bc = 1$. So the three roots must be $1, b, \frac{1}{b}$. We verify that as long as this holds, $P(x) = Q(x)$. It remains to find the minimum possible value of $1 + b + \frac{1}{b}$. Since $b + \frac{1}{b}$ is minimized for $b = 1$, the minimum value is $1 + 1 + 1 = 3$.

5. *(2021 P1, Solved By: 57%, Average Score: 4.1, Author: Daniel South)*

   There is exactly one possible triple:

   $$(x, b, c) = (-1, 2, 1).$$

   If x is even then $x^2 + bx + c \equiv 0 + 0 + 1 = 1 \pmod{2}$, so it cannot be a root. If x is odd, then $x^2 + bx + c \equiv 1 + b + 1 \equiv b \pmod{2}$, so it cannot be a root unless b is even. Therefore in order for the polynomial to have an integer root, b must be an even prime and thus equal to 2.

   Next we write $x^2 + 2x + c = (x + 1)^2 + (c - 1)$, which is never zero for $c \geq 2$, so $c = 1$ and $x = -1$. Thus the only solution is $(-1, 2, 1)$.

6. *(2019 P2, Solved By: 52%, Average Score: 3.8, Author: Grant Molnar)*

   There are three such polynomials: $p(x) = 0$, $p(x) = 1$, and $p(x) = x^2$.

   Solution 1: Let n be the degree of the polynomial p. Then the degree of $p(x)^2$ is $2n$ and the degree of $p(p(x))$ is n^2. Since $p(x)^2 = p(p(x))$, we conclude $2n = n^2$. Therefore, $n = 0$ or $n = 2$.

   - If $n = 0$, then let $p(x) = c$ (a constant). We have $c^2 = c$, so $c = 0$ or $c = 1$. This gives the first two solutions, $p(x) = 0$ and $p(x) = 1$.
   - If $n = 2$, say $p(x) = ax^2 + bx + c$, then the leading coefficient of $p(x)^2$ is a^2x^4, while the leading coefficient of $p(p(x))$ is a^3x^4. Thus

$a^2 = a^3$. Since $p(x)$ has degree 2, we know $a \neq 0$, hence $a = 1$. Therefore, we can write

$$p(p(x)) = p(x)^2 + bp(x) + c.$$

If this is equal to $p(x)^2$, then $p(x)^2 = p(x)^2 + bp(x) + c$, or $bp(x) + c = 0$. Since $p(x)$ has degree 2, the only way this can be true for all values of x is if $b = 0$, and then $c = 0$. Therefore, $p(x) = x^2$.

Solution 2: If $p(x)$ is constant, then $p(x) = 0$ or $p(x) = 1$. Otherwise, let $u = p(x)$, and the equation becomes $u^2 = p(u)$. Since $p(x)$ is not constant, $u = p(x)$ has infinitely many possible values. This implies that $y^2 = p(y)$ has infinitely many solutions for y (plug in $u = p(x)$ for y). But then the polynomial $p(y) - y^2$ has infinitely many roots, so it must be the zero polynomial, so $p(y) = y^2$. Renaming y to x, $p(x) = x^2$ for all x.

Note: While proposed independently, this problem is similar to the 1975 Canadian Mathematical Olympiad #8 [16].

7. *(2022 P3, Solved By: 29%, Average Score: 3, Author: Grant Molnar)*

There are two possible sequences: $a_n = -\frac{n^2}{2}$ or $a_n = \frac{n - n^2}{2}$.

Multiplying the second equation by 2 and adding $m^2 n^2$ to both sides, we find

$$2a_{mn} + (mn)^2 = (2a_m + m^2)(2a_n + n^2).$$

Therefore, define $b_n = 2a_n + n^2$, so the second equation can be written as $b_{mn} = b_m b_n$. Multiplying the first equation by 2 and adding $(m + n)^2$, we find

$$2a_{m+n} + (m + n)^2 = (2a_m + m^2) + (2a_n + n^2).$$

Therefore, $b_{m+n} = b_m + b_n$. Therefore, it suffices to find sequences b_1, b_2, b_3, ... that satisfy both $b_{mn} = b_m b_n$ and $b_{m+n} = b_m + b_n$.

We claim that if $b_{m+n} = b_m + b_n$ for all positive integers m, n, then $b_n = nb_1$. Certainly this is true for $n = 1$. If $b_k = kb_1$, then $b_{k+1} = b_k + b_1 = kb_1 + b_1 = (k + 1)b_1$, so by induction, it follows that $b_n = nb_1$ for all positive integers n. Additionally, we observe that any sequence of this form will satisfy $b_{m+n} = b_m + b_n$.

Substituting $m = n = 1$ into $b_{mn} = b_m b_n$, we find $b_1 = b_1^2$, so $b_1 = 0$ or $b_1 = 1$. Combining this with the previous paragraph, the only possible sequences are $b_n = 0$ or $b_n = n$. Indeed, both sequences satisfy both $b_{m+n} = b_m + b_n$ and $b_{mn} = b_m b_n$.

Therefore, since $a_n = \frac{b_n - n^2}{2}$, it follows that the two possible sequences are $a_n = -\frac{n^2}{2}$ or $a_n = \frac{n - n^2}{2}$.

Note: The sequence in this problem is the 2-derivation on the integers, and is a special case of Definition 2.1 of Alexandru Buium's monograph *Arithmetic Differential Equations* [4].

8. *(2021 P3, Solved By: 38%, Average Score: 2.8, Author: Grant Molnar)*

Solution 1: It is impossible to assign a number to each point in the plane such that the property is true. Given *any* two distinct points A and B in the plane, let M be the midpoint of $\overline{AB}$. We can construct points C and D on the perpendicular bisector of $\overline{AB}$, such that $CM = MD = \frac{AM}{\sqrt{3}}$. Note that $\triangle ACD$ and $\triangle BCD$ are congruent equilateral triangles.

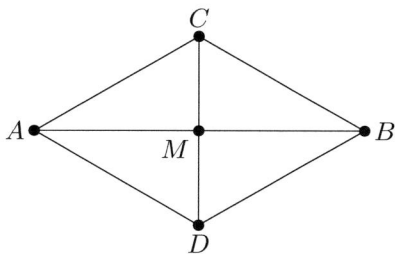

If such a function exists, then as $\triangle ACD$ and $\triangle BCD$ have the same perimeter, we know that

$$f(A) + f(C) + f(D) = f(B) + f(C) + f(D).$$

Hence $f(A) = f(B)$ for all distinct points A and B, so f must be equal to a constant for all points in the plane. If $f(P) = k$ for all points P in the plane, then it follows that $f(P) + f(Q) + f(R) = 3k$ for every equilateral triangle PQR. Hence the perimeter of every equilateral triangle in the plane is $3k$, which is a contradiction. We conclude that the requested task is impossible.

Solution 2: Suppose by way of contradiction that it is possible to make such an assignment. Let PQR be an equilateral triangle with side length 2, and let A, B, and C be the midpoints of $\overline{PQ}$, $\overline{QR}$, and $\overline{RP}$, respectively.

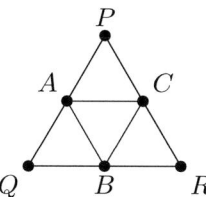

Note that $\triangle PAC$, $\triangle QBA$, $\triangle RCB$, and $\triangle ABC$ are equilateral triangles with side length 1. Then $f(P) + f(Q) + f(R) = 6$ and

$$(f(P) + f(A) + f(C)) = 3$$
$$(f(Q) + f(B) + f(A)) = 3$$
$$(f(R) + f(C) + f(B)) = 3$$
$$(f(A) + f(B) + f(C)) = 3$$

Adding these equations, and subtracting $f(P) + f(Q) + f(R) = 6$, we find

$$3\left(f(A) + f(B) + f(C)\right) = 6.$$

which implies that $f(A) + f(B) + f(C) = 2$. But the perimeter of $\triangle ABC$ is 3, so

$$f(A) + f(B) + f(C) = 3,$$

and we have a contradiction. Thus no such assignment exists.

9. *(2015 P4, Solved By: 14%, Average Score: 2.7, Author: Benjamin Stanford)*

We claim they can do it in a minimum of 16 hours.

Firstly, we show this is possible in the following way. For the first 6 hours, Balthazar takes the bike and goes 60 km. Then, Balthazar drops off the bike and walks for the remaining 10 hours, in which time he goes $4 \cdot 10 = 40$ km and thus reaches the destination 100 km away. Meanwhile, Anastasia walks for the first 12 hours, in which time she walks $12 \cdot 5 = 60$ km and arrives at the bike Balthazar dropped off. She then bikes for the remaining 4 hours, in which time she covers the remaining 40 km.

Now we show that no matter how they do this, they will take at least 16 hours. We may assume that Anastasia and Balthazar only stop, turn around, switch from biking to walking, or vice versa at a finite number of locations; fix these locations and suppose that the paths between locations are represented by line segments $1, 2, 3, \ldots, n$, with the length of line segment i being a_i, and with $a_1 + a_2 + a_3 + \cdots + a_n = 100$ km.

Fix a specific line segment i. Let x be the number of times a person bikes forward across it, x' be the number of times a person bikes backward across it, y be the number of times a person walks forward across it, and y' be the number of times a person walks backward across it. Since both Anastasia and Balthazar eventually get to the other side, $x + y - x' - y' = 2$. Since the bike must end up one side or the other of the segment, $x - x' = 0$ or 1. Thus $y - y' = 1$ or 2, and in particular $y \geq 1$. Thus either Anastasia walks forward across this line segment at some point, or Balthazar does (or both).

Let A be the total length Anastasia walks forward, and let B be the total length Balthazar walks forward. Since either Anastasia or Balthazar walks across each segment, $A + B \geq 100$ km. The total time spent walking by Anastasia is at least $A/5$ and the total time spent walking by Balthazar is at least $B/4$. Additionally, Anastasia spends at least $(100 - A)/10$ hours biking and Balthazar spends at least $(100 - B)/10$ hours biking. Thus if the total time it takes them is T, we have

$$T \geq \frac{A}{5} + \frac{100 - A}{10} = 10 + \frac{1}{10}A$$
$$T \geq \frac{B}{4} + \frac{100 - B}{10} = 10 + \frac{3}{20}B.$$

Adding 30 times the first inequality to 20 times the second inequality, we find

$$30T + 20T \geq 300 + 200 + 3A + 3B = 500 + 3(A + B).$$

Employing our observation earlier that $A + B \geq 100$, we find that

$$50T \geq 500 + 3(100) = 800,$$

so

$$T \geq \frac{800}{50} = 16 \text{ hours}.$$

10. *(2020 P2, Solved By: 16%, Average Score: 1.5, Author: Daniel South)*

First, if $a^n = b^n$ for any odd n, then $a = b$, and we are done. So assume that $a^n \neq b^n$ for all odd n. Then for all odd n,

$$\frac{a^{2n} - b^{2n}}{a^n - b^n} = a^n + b^n$$

is the ratio of two rational numbers, and so is rational. This implies that a^n and b^n are rational, since

$$a^n = \frac{(a^n + b^n) + (a^n - b^n)}{2}$$
$$\text{and} \qquad b^n = \frac{(a^n + b^n) - (a^n - b^n)}{2}.$$

In summary, for all odd n we have that a^n and b^n are rational. Finally, since a^5 and a^3 are rational, $a = \frac{(a^3)^2}{a^5}$ must be rational (or $a = 0$, which is also rational). An identical argument holds for b. Thus, a and b are rational.

Note: It is not necessary to use all $n \geq 2$; for example, the above solution only really uses $n = 3, 5, 6$, and 10. There are other combinations of n that also work. In fact, $n = 3, 5$, and 6 is enough, but the proof is significantly more difficult. In general, it is possible to show that we need at least two odd values of n and at least one even value of n in order to obtain a solution.

11. *(2016 P5, Solved By: 3%, Average Score: 0.3, Author: Samuel Dittmer)*

There are only two possible sequences: the constant sequence $1, 1, 1, \ldots$ and the constant sequence $-1, -1, -1, \ldots$.

Solution 1: We can factor the equation as

$$[a_{n+k} - ka_n][a_{n+k} - (k+1)a_n] = k(k-1). \tag{1}$$

Substituting in $k = 1$, we find that either $a_{n+1} = 2a_n$ or $a_{n+1} = a_n$ for all integers n.

If p is a prime such that $p \mid a_r$, then by repeatedly applying $a_{n+1} = 2a_n$ or $a_{n+1} = a_n$, we find that p divides a_n for all integers $n \geq r$. Substituting $k = 2$ and $n = r$ into (1) we notice that the left-hand side is divisible by p^2, and the right-hand side is equal to 2; hence $p^2 \mid 2$, which is a contradiction. Thus a_n is never divisible by any prime for any n, so $a_n = \pm 1$.

In particular, $a_{n+1} = a_n$ for all integers n, and a_n is constant—either 1 or -1. Substituting either of these sequences into (1) shows that both are valid solutions.

Solution 2: After multiplying by 4 and completing the square twice, we find

$$(2a_{n+k} - (2k+1)a_n)^2 - (2k-1)^2 = a_n^2 - 1.$$

As the closest square to $(2k-1)^2$ is $(2k-2)^2$ (for k positive), we see that if $x \neq 2k-1$ is an integer, then $|x^2 - (2k-1)^2| \geq 4k-3$. Thus if $|a_n^2 - 1| \neq 0$, then $|a_n^2 - 1| \geq 4k - 3$ for all positive integers k. But this contradicts the fact that $|a_n^2 - 1|$ is finite. Therefore, $a_n^2 = 1$ and $|2a_{n+k} - (2k+1)a_n| = 2k - 1$. Now if $a_{n+k} = -a_n$, then $|2a_{n+k} - (2k+1)a_n| = 2k + 3$, which is not equal to $2k - 1$, hence $a_{n+k} \neq -a_n$. But as $a_n^2 = 1$ for all n, we see that either $a_n = 1$ for all n, or $a_n = -1$ for all n. It is easy to check that both of these sequences work.

Solution 3: Factoring the equation as in (1) with $k = 1$, we find that either $a_{n+1} = a_n$ or $a_{n+1} = 2a_n$ for all nonnegative integers n. It follows that, for all nonnegative integers n, either (i) $a_{n+2} = a_n$, or (ii) $a_{n+2} = 2a_n$, or (iii) $a_{n+2} = 4a_n$. Now factor the equation with $k = 2$, and we get

$$(a_{n+2} - 2a_n)(a_{n+2} - 3a_n) = 2.$$

Fix $n \geq 0$. Consider the cases (i), (ii), and (iii). In case (ii), the above gives $0 = 2$, a contradiction. In case (i), the above gives $(-a_n)(-2a_n) = 2$, so $a_n^2 = 1$. In case (iii), $(2a_n)(a_n) = 2$, so $a_n^2 = 1$. In any case, $a_n^2 = 1$, so $a_n \in \{-1, 1\}$ for all n.

Because a_{n+1} has to be either a_n or $2a_n$, and $a_n \in \{-1, 1\}$ for all n, a_n must be constant. Thus the only sequences that satisfy the desired property are $a_n = 1$ for all $n \geq 0$, and $a_n = -1$ for all $n \geq 0$.

12. *(2014 P5, Solved By: 0%, Average Score: 0.2, Author: Hiram Golze)*

Ignore the first equation for now. From the second equation $x^2 + y^2 + z^2 = 4x\sqrt{yz} - 2yz$, we deduce

$$(x^2 - 4x\sqrt{yz} + 4yz) + (y^2 - 2yz + z^2) = 0$$
$$(x - 2\sqrt{yz})^2 + (y - z)^2 = 0$$

Since x, y, z are real, and as the sum of two squares can only be 0 when both squares are 0 (as squares are nonnegative), we conclude that both $y = z$ and

$$x - 2\sqrt{yz} = 0 \implies x = 2\sqrt{zz} = 2|z| = 2z.$$

Now we recall the first equation.

$$xyz = 1 \implies (2z)(z)(z) = 1 \implies z^3 = \frac{1}{2} \implies z = \frac{1}{\sqrt[3]{2}}.$$

Therefore, the only solution is:

$$x = \frac{2}{\sqrt[3]{2}} \ , \ y = \frac{1}{\sqrt[3]{2}} \ , \ z = \frac{1}{\sqrt[3]{2}}.$$

3.5 NUMBER THEORY

1. *(2013 Flyer, Author: Unknown)*

Notice that $f(n+1) - f(n) = 1$ if and only if there is exactly one power of 2 between 3^n and 3^{n+1}. Thinking multiplicatively, since $3 > 2$, there is always at least one power of 2 between them, and since $3 < 2^2 = 4$, there is always at most two between them. Therefore, $f(n+1) - f(n) \in \{1, 2\}$ for all n. Additionally, we have $f(0) = 0$, and

$$f(n) = (f(1) - f(0)) + (f(2) - f(1)) + \cdots + (f(n) - f(n-1)).$$

So $f(n)$ is the sum of n numbers, each of which is 1 or 2. If only finitely many of these are 1 for all n, then $f(n) \geq 2n - C$, where C is a constant. However, by the definition of f, $2^{f(n)} \leq 3^n$. It follows that $f(n) \leq n \log_2(3)$ for all n. Since $\log_2 3 < 2$, the sequence $2n - C$ grows faster than $n \log_2 3$, this is impossible.

2. *(2015 P3, Solved By: 45%, Average Score: 3.5, Author: Samuel Dittmer)*

The only possible values are $n = 3$ and $n = 9$.

Solution 1: By the Binomial Theorem,

$$(x+y)^n - x^n - y^n = \binom{n}{1}x^{n-1}y + \binom{n}{2}x^{n-2}y^2 + \cdots + \binom{n}{n-1}xy^{n-1}.$$

(This is the nth row of Pascal's triangle, without the initial and final 1s.) Therefore, if $n \geq 2$, then we must have $3 \mid \binom{n}{1}$, which is the same as $3 \mid n$. If $n \geq 4$, then we must have $3 \mid \binom{n}{3}$, i.e.,

$$3 \quad \Big| \quad \frac{n(n-1)(n-2)}{6}.$$

As n is divisible by 3, we know that $(n-1)$ and $(n-2)$ are not divisible by 3. Therefore, the only factors of 3 in the numerator can come from n. We need one factor of 3 to cancel with the factor of 3 in the denominator, and one factor to make the integer divisible by 3. Therefore, $9 \mid n$. If $n \geq 10$, then we must have $3 \mid \binom{n}{9}$. By a similar argument, using the expansion

$$3 \quad \Big| \quad \frac{n(n-1)(n-2)(n-3)(n-4)(n-5)(n-6)(n-7)(n-8)}{9 \cdot 8 \cdot 7 \cdot 6 \cdot 5 \cdot 4 \cdot 3 \cdot 2}$$

all the factors of 3 all come from n, $n-3$, $n-6$ on the top and $9, 6$, and 3 on the bottom; since $n-3$ and $n-6$ only contain one factor of 3, we conclude $27 \mid n$. But this is impossible since $n \leq 20$.

Thus, only numbers to check are $n = 3, 9$. In fact, both of them satisfy the desired property.

Solution 2: If the greatest common divisor (gcd) of the coefficients is divisible by 3, then the sum of the coefficients is divisible by 3, hence $(1+1)^n - (1+1) = 2^n - 2$ is divisible by 3. But as $2^2 \equiv 1 \pmod 3$, we see that

$$2^n \equiv \begin{cases} 1 \pmod 3 & \text{if } n \text{ is even} \\ 2 \pmod 3 & \text{if } n \text{ is odd.} \end{cases}$$

Hence $2^n - 2$ is divisible by 3 if and only if n is odd. Also, the coefficient of $x^{n-1}y$ is n, so n must be divisible by 3. Hence n must be congruent to 3 (mod 6), leaving $n = 3, 9, 15$. We check that $n = 3, 9$ satisfy the property, and for $n = 15$, we find that $\binom{15}{3}$ is not divisible by 3, so 3 and 9 are the only values that work.

Note: In general, the greatest common divisor of the coefficients of $(x+y)^n - x^n - y^n$ is 3 if and only if n is a perfect power of 3 ($n = 3, 9, 27, 81$, and so on).

3. *(2022 P2, Solved By: 47%, Average Score: 3.4, Author: Grant Molnar)*

   Solution 1: The problem is equivalent to showing there are no primes p dividing both $x^2 + xy + y^2$ and $x^2 + 3xy + y^2$. If p is such a prime, then p divides their difference, which is $2xy$. Thus $p = 2$, p divides x, or p divides y. Without loss of generality, $p = 2$ or p divides x.

   - If $p = 2$, then $x^2 + xy + y^2$ must be even. However, since x and y are coprime, they cannot both be even. If both x and y are odd, then $x^2 + xy + y^2$ is odd, and if one of x, y is even and the other is odd, then $x^2 + xy + y^2$ is also odd. Therefore, it is impossible for $x^2 + xy + y^2$ to be even, and we obtain a contradiction.

- If p divides x, then since p divides $x^2 + xy + y^2$ and $x^2 + xy$, p divides the difference y^2. Since p is prime, it follows that p divides y. But this contradicts the fact that x and y are coprime.

Since both cases lead to a contradiction, no such prime p exists.

Solution 2: We need to show that the greatest common divisor (gcd) of $x^2 + xy + y^2$ and $x^2 + 3xy + y^2$ is 1, given that $\gcd(x, y) = 1$. To do this, we use the following well-known rules for computing the greatest common divisor:

(S) Symmetry: $\gcd(a, b) = \gcd(b, a)$

(E) Euclidean algorithm: $\gcd(a, b) = \gcd(a, b - ka)$ for any $k \in \mathbb{Z}$

(P) Product rule: $\gcd(a, bc) \mid \gcd(a, b) \cdot \gcd(a, c)$

Applying these rules, we find, where $F = x^2 + xy + y^2$:

$$\begin{aligned}
\gcd(F&, x^2 + 3xy + y^2) \\
&= \gcd(F, 2xy) && \text{by (E)} \\
&\mid \gcd(F, 2) \cdot \gcd(F, x) \cdot \gcd(F, y) && \text{by (P)} \\
&= \gcd(F, 2) \cdot \gcd(x, F) \cdot \gcd(y, F) && \text{by (S)} \\
&= \gcd(F, 2) \cdot \gcd(x, y^2) \cdot \gcd(y, x^2) && \text{by (E)} \\
&\mid \gcd(F, 2) \cdot \gcd(x, y)^2 \cdot \gcd(y, x)^2 && \text{by (P)} \\
&= \gcd(F, 2) \cdot \gcd(x, y)^4 && \text{by (S)} \\
&= \gcd(F, 2) && \text{since } \gcd(x, y) = 1 \\
&= 1,
\end{aligned}$$

because $x^2 + xy + y^2$ is always odd for coprime x, y. (This can be seen by cases on whether x and y are odd or even.) Since the gcd of the two polynomials divides 1, it must be 1.

4. *(2015 P1, Solved By: 18%, Average Score: 2.5, Author: Hiram Golze, Samuel Dittmer, and Wyatt Mackey)*

Suppose that the three trolls have a, b, and c pancakes, respectively. We claim that $22 = 4 + 6 + 12$ is the smallest possible sum of a, b, and c.

If we choose $a = 4$, $b = 6$, and $c = 12$, the three properties are satisfied, and $a + b + c = 22$. So it remains to show that 22 is minimal.

We first claim that none of a, b, and c can be prime. If (without loss of generality) a were prime, then a would be divisible by both of the greatest common divisors $\gcd(a, b)$ and $\gcd(a, c)$, which are distinct numbers greater than 1, contradicting the fact that a is prime. Therefore, we may assume that a, b, and c are composite. The first few composite numbers are $4, 6, 8, 9, 10, 12, \ldots$.

We next note that if two of the numbers are equal, say $a = b$, then the second condition is violated because $\gcd(a, c) = \gcd(b, c)$. Therefore, the three numbers are distinct.

If none of the numbers equal 4, then the sum must be greater than or equal to $6 + 8 + 9 = 23$, but we can already do better than that. Therefore, for a minimal n, we must have one of the numbers (say a) equal to 4. $\gcd(4, b)$ and $\gcd(4, c)$ are both distinct factors of 4 greater than 1. Without loss of generality, they must be 2 and 4, respectively. So b is an odd multiple of 2 greater than 4, and c is a multiple of 4 greater than 4. So $\gcd(b, c)$ must be a multiple of 2 greater than 4, so it is at least 6. Then b, c are at least the smallest two multiples of 6, so the total sum is at least $4 + 6 + 12 = 22$, as desired.

5. *(2019 P3, Solved By: 30%, Average Score: 2.4, Author: Caleb Stanford)*

There are five possible triples (a, b, c):

$$(0, 0, 2), (1, 2, 3), (2, 1, 3), (4, 3, 4), (3, 4, 4).$$

Solution 1: For $c \leq 4$, we write $c!$ in binary:

$$0! = 1! = 1 = 1_2$$
$$2! = 2 = 10_2$$
$$3! = 6 = 110_2$$
$$4! = 24 = 11000_2$$

Here, N_2 denotes that N is the binary representation of an integer. Since binary representation is unique, we see that only $3!$ and $4!$ can be written as a sum of two *distinct* powers of two, and each in two different ways (reordering the two powers). This gives the four solutions $(1, 2, 3)$, $(2, 1, 3)$, $(4, 3, 4)$, and $(3, 4, 4)$. Also, only $2!$ can be written as a sum of two of the *same* power of two, $2! = 2^0 + 2^0$. This gives the solution $(0, 0, 2)$. These are the only possible solutions for $c \leq 4$.

Now we assume that $c \geq 5$. Thus, $c!$ is divisible by both 3 and 5. Also, we may assume without loss of generality that $a \leq b$, and then

$$2^a + 2^b = 2^a(1 + 2^{b-a}).$$

Therefore, if $2^a + 2^b = c!$, then $1 + 2^{b-a}$ must be divisible by 15. But we can list the powers of 2 (mod 15), finding

$$1, 2, 4, 8, 1, 2, 4, 8, \ldots.$$

Therefore, powers of 2 are periodic mod 15 with period 4. In particular, $2^n \not\equiv -1 \pmod{15}$ for any n. Therefore, $1 + 2^{b-a}$ cannot be divisible by 15, and there are no more solutions.

Solution 2: For any $k \geq 0$, we have $2^k \equiv 1, 2$, or $4 \pmod 7$. As you cannot add $1, 2$, or 4 to $1, 2$, or 4 to get 0 $\pmod 7$, it follows that $2^a + 2^b$ is not a multiple of 7 for any a, b. However, since $7 \mid c!$ for $c \geq 7$, it follows that $c \leq 6$.

Once we know $c \leq 6$, we proceed as in Solution 1 to write $c!$ in binary. For $c = 0, 1, 2, 3, 4$, we get the same set of solutions as before. For $c = 5$, we get $5! = 120 = 1111000_2$, which contains four 1s, so cannot be a sum of two powers of 2. Finally, for $c = 6$, we get $6! = 720 = 1011010000_2$, which also cannot be a sum of two powers of 2.

Note: While proposed independently, this problem is identical to problem #8 from the 2003 Harvard/MIT Math Tournament Guts round [20].

6. *(2014 P2, Solved By: 19%, Average Score: 2.4, Author: Hiram Golze)*

Solution 1: We present a solution that works for both parts (a) and (b).

There are no solutions to the equations in question. Consider the equations mod 8. They become the same congruence.

$$x^2 + y^2 \equiv 6 \pmod 8.$$

If the equations had a solution, then this congruence would also have a solution, so it suffices to show that this congruence has no solutions. The only perfect squares mod 8 are 0, 1, and 4. Therefore, the only possibilities for the sums of squares are $0, 1, 2, 4, 5 \pmod 8$. Therefore, the sum of two squares is never congruent to 6 $\pmod 8$. Hence the equations have no solutions.

The same argument can be adapted for part (b) of the problem.

Solution 2: Suppose that the first equation has some solution (x, y). If x is even, then x^2 will be even, and if x is odd, then x^2 will be odd. Therefore, because the sum of an even number and an odd number is odd, we know that either x and y are both even, or else they are both odd. If x and y are both even, let $x = 2x'$ and $y = 2y'$ for integers x' and y'. Then the first equation becomes

$$4(x')^2 + 4(y')^2 = 2014.$$

But $2014 = 2 \cdot 1007$, so it is not divisible by 4, so this case is impossible. Therefore, x and y must both be odd. Therefore, we can represent them by $x = 2x' - 1$ and $y = 2y' - 1$ for integers x' and y'. Then the first equation becomes

$$(4(x')^2 - 4x' + 1) + 4(y')^2 - 4y' + 1) = 2014.$$
$$4((x')^2 - x' + (y')^2 - y') = 2012$$
$$(x')^2 - x' + (y')^2 - y' = 503.$$

Now $(x')^2 - x = x'(x' - 1)$ is the product of two consecutive integers, and so one of these consecutive integers must be even. Therefore, the product must be even. The same applies to $(y')^2 - y'$, hence their sum must be even. But the sum is 503, which is odd, so we have a contradiction. Therefore, the first equation has no solutions.

The same argument can be adapted for part (b) of the problem.

7. *(2020 P5, Solved By: 11%, Average Score: 1.1, Author: Hiram Golze)*

   Solution 1: By the Law of Cosines and its converse, a triangle with side lengths a, b, c will have an angle of $120°$ between the sides with lengths a and b if and only if

   $$c^2 = a^2 + b^2 - 2ab \cos 120°$$
   $$= a^2 + b^2 + ab.$$

   For any $m \geq 1$, set $a = 2m + 1$, $b = 3m^2 + 2m$ and $c = 3m^2 + 3m + 1$. Note that $a < b < c$ and $a + b > c$, so this is a valid triangle. Also, we find that

   $$a^2 + b^2 + ab$$
   $$= (2m + 1)^2 + (3m^2 + 2m)^2 + (2m + 1)(3m^2 + 2m)$$
   $$= (4m^2 + 4m + 1) + (9m^4 + 12m^3 + 4m^2) + (6m^3 + 7m^2 + 2m)$$
   $$= 9m^4 + 18m^3 + 15m^2 + 6m + 1$$
   $$= (3m^2 + 3m + 1)^2$$
   $$= c^2,$$

   so this triangle is special.

   Then, using the Euclidean algorithm we can compute the greatest common divisor

   $$\gcd(b, c) = \gcd(3m^2 + 2m, m + 1) = \gcd(-m, m + 1) = 1.$$

   Therefore, b and c share no common factor greater than 1, and so this triangle is primitive.

   Note: This solution may seem like we pulled it out of a hat. However, there is some intuition behind it. It might make sense to see if we can generate a triple starting with an odd number $a = 2m + 1$ (much like there is an infinite family of Pythagorean triples $(2m+1, 2m^2+2m, 2m^2+ 2m + 1)$ that generates $(3, 4, 5)$, $(5, 12, 13)$, $(7, 24, 25)$, etc.). If there is a special triangle with $a = 2m + 1$, then

   $$c^2 = (2m + 1)^2 + b^2 + (2m + 1)b.$$

Multiplying by 4 and completing the square, we find

$$(2c)^2 = 3(2m+1)^2 + (2b + (2m+1))^2.$$

Therefore, the difference between squares $(2c)^2$ and $(2b + (2m+1))^2$ is $3(2m+1)^2$, which is an odd number. In fact, since each odd number is the difference between consecutive squares, we find that

$$3(2m+1)^2 = (6m^2 + 6m + 2)^2 - (6m^2 + 6m + 1)^2.$$

Setting $2c = 6m^2 + 6m + 2$ and $2b + (2m+1) = 6m^2 + 6m + 1$, we obtain the triple that we initially stated.

Solution 2: Assume that $b = a + 2$ and that the angle between a and b is $120°$. We will show infinitely many solutions in this case. By the Law of Cosines, the triangle is special if and only if

$$(a+2)^2 + a^2 + a(a+2) = c^2$$

which rearranges to
$$c^2 - 3(a+1)^2 = 1.$$

This is Pell's equation $x^2 - 3y^2 = 1$, so it has infinitely many solutions. For any such solution, set $c = x$ and $a = y - 1$ to get a special triangle (so $b = y + 1$).

For such a triangle, $\gcd(a, b, c)$ is at most 2, since $\gcd(a, b) = \gcd(y - 1, y + 1) = \gcd(y - 1, 2)$ divides 2. Since there are infinitely many solutions to Pell's equation, there are two cases: either there are infinitely many special a, b, c where $\gcd(a, b, c) = 1$, or there are infinitely many special a, b, c where $\gcd(a, b, c) = 2$. In the former case, we have infinitely many primitive special triangles and we are done. In the latter case, take the infinitely many special triangles and divide each side length by 2. This results in infinitely many distinct special triangles a', b', c' where $\gcd(a', b', c') = 1$.

8. *(2013 P3, Solved By: 7%, Average Score: 1, Author: Samuel Dittmer)*

   We claim that the only possibilities for x are 38, 538, 462, and 962.

   Solution 1: Let $x = 100a + 10b + c$, with $a, b, c \in \{0, 1, 2, ..., 9\}$. We are looking for when the last three digits of x^2 are ddd, with $d \in \{1, 2, ..., 9\}$. Now the problem becomes finding when

   $$(100a + 10b + c)^2 \equiv 111d \pmod{1000}.$$

   Expanding out $(100a + 10b + c)^2$ this becomes

   $$200ac + 100b^2 + 20bc + c^2 \equiv 100d + 10d + d \pmod{1000} \qquad (1)$$

   Since 20 divides 1000, we can take this equation modulo 20 to simplify things. Doing this, it becomes $c^2 \equiv 11d \pmod{20}$. Now consider all possible cases for c and d.

c	$c^2 \pmod{20}$
0	0
1	1
2	4
3	9
4	16
5	$25 \equiv 5$
6	$36 \equiv 16$
7	$49 \equiv 9$
8	$64 \equiv 4$
9	$81 \equiv 1$

d	$11d \pmod{20}$
1	11
2	$22 \equiv 2$
3	$33 \equiv 13$
4	$44 \equiv 4$
5	$55 \equiv 15$
6	$66 \equiv 6$
7	$77 \equiv 17$
8	$88 \equiv 8$
9	$99 \equiv 19$

As we can see above, the only possibility for when $c^2 \equiv 11d \pmod{20}$ is if $c^2 \equiv 11d \equiv 4 \pmod{20}$, which implies $c = 2$ or 8 and $d = 4$. We split into the cases $c = 2$ and $c = 8$.

Case 1: $c = 2$.

Return to equation (1) and plug in the new values for c and d:

$$400a + 100b^2 + 40b + 4 \equiv 444 \pmod{1000}$$

$$400a + 100b^2 + 40b \equiv 440 \pmod{1000}$$

Now divide by the common factor of 20 to get:

$$20a + 5b^2 + 2b \equiv 22 \pmod{50} \qquad (2)$$

This implies $5b^2 + 2b \equiv 2 \pmod{10}$. Modulo 2 we find $5b^2 \equiv 0$, so b must be even; modulo 5 we must have $2b \equiv 2$ so $b \equiv 1 \pmod{5}$. By the Chinese Remainder Theorem, $b \equiv 0 \pmod{2}$ and $b \equiv 1 \pmod{5}$ imply $b \equiv 6 \pmod{10}$, so $b = 6$.

Now returning to (2) we get

$$20a + 5b^2 + 2b = 20a + 5 \cdot 36 + 2 \cdot 6 \equiv 22 \pmod{50}$$

$$\Rightarrow 20a + 20 \equiv 0 \pmod{50}$$

$$\Rightarrow 2a + 2 \equiv 0 \pmod{5}$$

$$\Rightarrow a \equiv 4 \pmod{5} \Rightarrow a = 4 \text{ or } a = 9$$

So we have the solutions 462 and 962, which we check both works.

Case 2: $c = 8$.

Return to equation (1) and plug in the new values for c and d:

$$1600a + 100b^2 + 160b + 64 \equiv 444 \pmod{1000}$$

$$1600a + 100b^2 + 160b \equiv 380 \pmod{1000}$$

Now divide by the common factor of 20 to get:

$$80a + 5b^2 + 8b \equiv 19 \quad (\text{mod } 50) \tag{3}$$

This implies $5b^2 + 8b \equiv 9$ (mod 10). We find $5b^2 \equiv 1$ (mod 2), so b must be odd; we must have $3b \equiv 4$ (mod 5) so $b \equiv 3$ (mod 5). By the Chinese Remainder Theorem, $b \equiv 1$ (mod 2) and $b \equiv 3$ (mod 5) imply $b \equiv 3$ (mod 10), so $b = 3$.

Now returning to (3) we get

$$80a + 5b^2 + 8b = 80a + 5 \cdot 9 + 8 \cdot 3 \equiv 19 \quad (\text{mod } 50)$$

$$\Rightarrow 80a \equiv 0 \quad (\text{mod } 50)$$

$$\Rightarrow 3a \equiv 0 \quad (\text{mod } 5)$$

$$\Rightarrow a \equiv 0 \quad (\text{mod } 5) \Rightarrow a = 0 \text{ or } a = 5$$

So we have the solutions 38 and 538, which we check both works.

Solution 2: We are attempting to solve $x^2 \equiv 111 \cdot y$ (mod 1000), for $1 \le y \le 9$. By (mod 10) considerations, we can immediately restrict to the cases $111, 444, 555, 666$, and 999. Now, as a consequence of the Chinese Remainder Theorem, the residue of n (mod 1000) is uniquely determined by the residue of n (mod 8) and (mod 125). The only perfect squares (mod 8) are $0, 1$, and 4. However, $111 \equiv 7$ (mod 8), $444 \equiv 4$ (mod 8), $555 \equiv 3$ (mod 8), $666 \equiv 2$ (mod 8), and $999 \equiv$ (mod 7), we can restrict to the case $y = 4$.

Now, $444 \equiv 4$ (mod 8) and $444 \equiv 69$ (mod 125), so we must solve $x_1^2 \equiv 4$ (mod 8) and $x_2^2 \equiv 69$ (mod 125). Certainly, we know $x_1 \equiv 2, 6$ (mod 8), i.e., $x_1 \equiv 2$ (mod 4). We solve the second equation first (mod 5), then (mod 25), then (mod 125):

- $x_2^2 \equiv 4$ (mod 5), so $x_2 \equiv 2, 3$ (mod 5).
- $(5x_2' \pm 2)^2 \equiv 19$ (mod 25), so $x_2' \equiv \pm 2$ (mod 5), and $x_2 \equiv \pm 12$ (mod 25). $(5x_2'' \pm 12)^2 \equiv 69$ (mod 125), so $x_2'' \equiv \pm 3$ (mod 5), and $x_2 \equiv \pm 87$ (mod 125) $= \pm 38$ (mod 125).

So then we have to simply find all x with $x \equiv 2$ (mod 4), and $x \equiv \pm 38$ (mod 125). This gives $38, 38 + 500 = 538, 87 + 375 = 462$, and $87 + 875 = 962$. Thus the only possibilities are $38, 462, 538$, and 962.

9. *(2017 P4, Solved By: 6%, Average Score: 0.8, Author: Hiram Golze)*

The minimum possible degree of P is 2.

Solution 1: Define $Q(x) = xP(x) - 1$. If $P(x)$ has degree k, then $Q(x)$ has degree $k + 1$, with a constant term of -1. Therefore, we wish to find the minimal degree (greater than or equal to 1) of $Q(x)$ such that $Q(x)$ has a constant term of -1 and $Q(n) \equiv 0$ (mod 16) for all odd n.

Taking

$$Q(x) = (x+1)^2(x-1), \tag{1}$$

then for all odd n, we see that $n+1$ and $n-1$ are even, and one of them is divisible by 4, hence $Q(n)$ is divisible by 16 for all odd n. Also, $Q(x)$ has a constant term of -1, so this $Q(x)$ works and has degree 3. Thus the minimum possible degree for Q is less than or equal to 3. It remains to rule out degree 1 and 2.

It is clear that $Q(x)$ cannot be linear, because if $Q(x) = ax - 1$, then $Q(1) = a-1 \equiv 0$ (mod 16) implies $a \equiv 1$ (mod 16), and $Q(3) = 3a-1 \equiv 0$ (mod 16) implies $a \equiv 11$ (mod 16), and these are contradictory.

If $Q(x)$ is quadratic, say $Q(x) = ax^2 + bx - 1$, then

$$Q(1) = a + b - 1 \equiv 0 \quad (\text{mod } 16)$$
$$\text{and} \quad Q(-1) = a - b - 1 \equiv 0 \quad (\text{mod } 16).$$

Subtracting these, we find that $2b \equiv 0$ (mod 16). Adding the same equations, we find that $2a \equiv 2$ (mod 16). Also, we find that

$$Q(3) = 9a + 3b - 1 \equiv 0 \quad (\text{mod } 16).$$

Subtracting the congruence for $Q(1)$, we find $8a + 2b \equiv 0$ (mod 16). But $2b \equiv 0$ (mod 16), hence $8a \equiv 0$ (mod 16). But as $2a \equiv 2$ (mod 16), we also know that $8a \equiv 8$ (mod 16). This is a contradiction, hence $Q(x)$ cannot be quadratic.

Therefore, the minimum possible degree for $Q(x)$ is 3, so the minimum possible degree for $P(x)$ is 2. Using the relation $Q(x) = xP(x) - 1$ to solve for $P(x)$ from $Q(x)$ in (1), we find that this occurs when

$$P(x) = \frac{(x+1)^2(x-1) + 1}{x} = x^2 + x - 1.$$

Solution 2: Note that if $nP(n) \equiv 1$ (mod 16), then $P(n) \equiv n^{-1}$ (mod 16), the multiplicative inverse of n mod 16. Therefore, we can compute the following values for $P(n)$ (mod 16).

n	1	3	5	7	9	11	13	15
$P(n)$	1	11	13	7	9	3	5	15

Using the method of finite differences, we know that if $P(x)$ has degree $k+1$, then $P(x+2) - P(x)$ has degree k. Therefore, we define the polynomial $P_0(x) = P(x)$, and in general, $P_{j+1}(x) = P_j(x+2) - P(x)$. If $P(x) = P_0(x)$ has degree $k+1$, then $P_{k+1}(x)$ must be constant, which implies that $P_{k+1}(x)$ (mod 16) must also be constant. Using the values of $P(n)$ from the table above, we can compute values of $P_j(x)$ (mod 16).

$P_0(n)$	1		11		13		7		9		3		5		15
$P_1(n)$		10		2		10		2		10		2		10	
$P_2(n)$			8		8		8		8		8		8		

As $P_1(x)$ is not constant mod 16, we see that it is impossible for P to have degree 1. But $P_2(x)$ is constant mod 16, so it looks like it might be possible for $P(x)$ to have degree 2. If $P(x)$ were to have degree 2, then $P_2(x)$ would be a constant integer, and from the above table it must be 8. Working backward, we can fill in the above table with integers rather than residues, finding P_1 from P_2 and P_0 from P_1:

$P_0(n)$	1		11		29
$P_1(n)$		10		18	
$P_2(n)$			8		

Then we wish to find $P(x)$ such that $P(1) = 1$, $P(3) = 11$, and $P(5) = 29$. If $P(x) = ax^2 + bx + c$, then we find

$$a + b + c = 1$$
$$9a + 3b + c = 11$$
$$25a + 5b + c = 29.$$

Solving this system of equations, we find $a = 1$, $b = 1$, and $c = -1$. This suggests that $P(x) = x^2 + x - 1$ is such a polynomial. Note that

$$n(n^2 + n - 1) = (n + 1)^2(n - 1) + 1,$$

so if n is odd, then $n+1$ and $n-1$ are even, and at least one of them is a multiple of 4. Hence $(n + 1)^2(n - 1)$ is divisible by 16. Hence $nP(n) \equiv 1$ (mod 16) for all odd n, so we have a satisfactory polynomial of degree 2. Therefore, the minimum possible degree of P is 2.

10. *(2021 P6, Solved By: 5%, Average Score: 0.6, Author: Caleb Stanford)*

Let the prime factorization of $n!$ be $p_1^{x_1} p_2^{x_2} p_3^{x_3} \cdots p_k^{x_k}$, where $p_1 = 2$, $p_2 = 3$, and so on are all the prime numbers between 1 and n, inclusive. The number of divisors of $n!$ is $(x_1 + 1)(x_2 + 1)(x_3 + 1) \cdots (x_k + 1)$.

We consider the following algorithm which assigns each value $x_i + 1$ with a distinct number between 1 and $2n$. Visit the prime numbers in order, starting with p_1 and ending with p_k. For each p_i, assign $(x_i + 1)$ to a multiple of $(x_i + 1)$ that has not yet been assigned. Assuming such a multiple always exists, in the end we have that each $(x_i + 1)$ divides the number it is assigned to, and the product of all assigned numbers divides $1 \cdot 2 \cdot 3 \cdots (2n - 1) \cdot (2n) = (2n)!$. Thus, it remains to show that at each step there is a multiple of $x_i + 1$ between 1 and $2n$ that is not assigned.

The number of factors of p_i going into $n!$ is given by the formula

$$x_i = \sum_{j=1}^{\infty} \left\lfloor \frac{n}{p_i^j} \right\rfloor,$$

(where $\lfloor \cdot \rfloor$ is the floor function), so it can be bounded above by a geometric series:

$$x_i < \sum_{j=1}^{\infty} \frac{n}{p_i^j} = n \frac{(1/p_i)}{1 - (1/p_i)} = \frac{n}{p_i - 1}.$$

Therefore,

$$x_i + 1 < \frac{n + p_i - 1}{p_i - 1} \le \frac{2n - 1}{p_i - 1} < \frac{2n}{p_i - 1},$$

which implies there are at least $p_i - 1$ multiples of $(x_i + 1)$ between 1 and $2n$. On the other hand, the number of primes so far (p_1 through p_{i-1}) is at most $p_i - 2$, since 1 is not prime. So this completes the proof that there is always a multiple of $(x_i + 1)$ available.

11. *(2016 P6, Solved By: 0%, Average Score: 0.3, Author: Caleb Stanford)*
    There are nine possible pairs (u, m):

    $$(1, 2), (1, 3), (2, 1), (2, 2), (2, 5), (3, 1), (3, 5), (5, 2), (5, 3).$$

    Solution 1: If $u + m^2$ is divisible by $um - 1$, then

    $$u(u + m^2) - m(um - 1) = u^2 + m$$

    must also be divisible by $um - 1$. Therefore, if (u, m) works, then (m, u) works, so we may assume that $m \le u$.
    As $(um - 1) \mid (u + m^2)$, we see that $um - 1 \le u + m^2$. Rearranging this, we find $u(m - 1) \le m^2 + 1$. Therefore, either $m = 1$ or

    $$u \le \frac{m^2 + 1}{m - 1} = m + 1 + \frac{2}{m - 1},$$

    which means u is quite close to m. In particular, if $m \ge 4$, then $u \le m + \frac{5}{3}$. But we assumed that $m \le u$, hence $u \le m + \frac{5}{3} \le u + \frac{5}{3}$. The only integers m that satisfy this are $m = u$ and $m = u - 1$. If $m = u$, then $u^2 - 1$ divides $u + u^2$, and these numbers have a common factor of $u + 1$, hence $u - 1$ divides u. But $u \ge m \ge 4$, so this is impossible. If $m = u - 1$, then $u(u - 1) - 1$ divides $u + (u - 1)^2$, or $u^2 - u - 1$ divides $u^2 - u + 1$. But when $u \ge 4$, $u^2 - u - 1 \ge 11$ (it's an increasing quadratic function). A number that is greater than 11 clearly cannot be a factor of a number that is two greater than that number. Therefore, this is impossible. Hence $m \le 3$. We enumerate each of these possibilities separately.

- If $m = 1$, then $u - 1$ divides $u + 1$. This is only possible if $u = 2, 3$.
- If $m = 2$, then $2u - 1$ divides $u + 4$. Note that $2u - 1 > u + 4$ if $u > 5$. Hence $u \leq 5$. Testing the five possible values of u, we find that $u = 1, 2, 5$ work.
- If $m = 3$, then $3u - 1$ divides $u + 9$. Note that $3u - 1 > u + 9$ if $u > 5$. Hence $u \leq 5$. Testing the five possible values of u, we find that $u = 1, 5$ work.

We assumed that $m \leq u$, so we also must include the reversed pairs. Doing so, we get the list of all solutions

$$(1, 2), (1, 3), (2, 1), (2, 2), (2, 5), (3, 1), (3, 5), (5, 2), (5, 3).$$

Solution 2: Suppose that $u + m^2 = (um - 1)q$ for some positive integer q. Then $m^2 - qum + u + q = 0$. By the quadratic formula,

$$m = \frac{qu \pm \sqrt{q^2 u^2 - 4(u + q)}}{2}. \tag{1}$$

In order for this to be an integer, the discriminant of the polynomial, $q^2 u^2 - 4(u + q)$, must be a perfect square. Note that $q^2 u^2$ is a perfect square, and the next smallest perfect square is $(qu - 1)^2 = q^2 u^2 - 2qu + 1$. For $u \geq 5$ and $q \geq 4$, we can squeeze $q^2 u^2 - 4(u + q)$ in between $(qu - 1)^2$ and $(qu)^2$.

To show this, note that if $u \geq 5$ and $q \geq 4$, then certainly $(q - 2)(u - 2) \geq 5$. Hence, we have

$$(q - 2)(u - 2) \geq 5$$
$$qu - 2u - 2q + 4 \geq 5$$
$$2qu - 4u - 4q \geq 2$$
$$q^2 u^2 - 4u - 4q \geq q^2 u^2 - 2qu + 2 = (qu - 1)^2 + 1.$$

Therefore, $(qu - 1)^2 < q^2 u^2 - 4u - 4q < (qu)^2$. Hence if $u \geq 5$ and $q \geq 4$, then $q^2 u^2 - 4u - 4q$ cannot be a perfect square, hence $um - 1$ cannot divide $u + m^2$. Therefore, either $u \leq 4$ or $q \leq 3$. We enumerate each of these cases below.

- If $u = 1$, then we want $(m - 1) \mid (1 + m^2)$. Clearly $(1, 1)$ does not work, while $(1, 2)$ and $(1, 3)$ satisfy the desired property. Otherwise, $1 + m^2 = (m - 1)(m + 1) + 2$, so if $1 + m^2$ is divisible by $m - 1$, then 2 is also divisible by $m - 1$. But this can only happen if $m - 1 = 1, 2$. Hence $m = 2, 3$ are indeed the only possibilities.
- If $u = 2$, then we want $(2m - 1) \mid (2 + m^2)$. Therefore, $(2m - 1) \mid (8 + 4m^2)$, and

$$4m^2 + 8 = (2m - 1)(2m + 1) + 9.$$

Hence $(2m - 1) \mid 9$. This is only possible if $2m - 1 = 1, 3, 9$, or $n = 1, 2, 5$. We see that $(2, 1)$, $(2, 2)$, $(2, 5)$ satisfy the desired property.

- If $u = 3$, then we want $(3m - 1) \mid (3 + m^2)$. Therefore,

$$(3m - 1) \mid (27 + 9m^2),$$

and
$$9m^2 + 27 = (3m - 1)(3m + 1) + 28.$$

Hence $(3m - 1) \mid 28$. This is only possible if

$$3m - 1 = 1, 2, 4, 7, 14, 28,$$

and the only integers m that satisfy one of these equations are $m = 1, 5$. We see that $(3, 1)$, $(3, 5)$ satisfy the desired property.

- If $u = 4$, then we want $(4m - 1) \mid (4 + m^2)$. Therefore, $(4m - 1) \mid (64 + 16m^2)$, and

$$16m^2 + 64 = (4m - 1)(4m + 1) + 65.$$

Hence $(4m - 1) \mid 65$. This is only possible if $4m - 1 = 1, 5, 13, 65$, and none of these yield integers m. Therefore, there are no solutions with $u = 4$.

- If $q = 1$, then the discriminant in (1) is $u^2 - 4u - 4$. This is clearly less than $u^2 - 4u + 4 = (u - 2)^2$. Also, if $2u \geq 14$, then $u^2 - 4u - 4 \geq u^2 - 6u + 10 = (u - 3)^2 + 1$. Therefore, if $u \geq 7$, then $u^2 - 4u - 4$ cannot be a perfect square. Checking through $u = 1, 2, \ldots, 6$, we find that $u^2 - 4u - 4$ is only a perfect square for $u = 5$, and then by (1), $m = \frac{5 \pm \sqrt{25 - 24}}{2} = 2$ or 3. We see that $(5, 2)$ and $(5, 3)$ satisfy the desired property.

- If $q = 2$, then the discriminant in (1) is $4u^2 - 4u - 8$. This is strictly less than $4u^2 - 4u + 1 = (2u - 1)^2$. Also, if $4u \geq 13$, then $4u^2 - 4u - 8 \geq 4u^2 - 8u + 5 = (2u - 2)^2 + 1$. Therefore, $4u^2 - 4u - 8$ is not a perfect square for $u \geq 4$. Checking through $u = 1, 2, 3$, we find that $4u^2 - 4u - 8$ is only a perfect square when $u = 2, 3$. If $u = 2$, then by (1), $m = \frac{4 \pm \sqrt{16 - 16}}{2} = 2$, while if $u = 3$, then by (1), $m = \frac{6 \pm \sqrt{36 - 20}}{2} = 1$ or 5. We see that $(2, 2)$, $(3, 1)$, and $(3, 5)$ satisfy the desired property.

- If $q = 3$, then the discriminant in (1) is $9u^2 - 4u - 12$. This is strictly less than the perfect square $(3u)^2$. Also, if $2u \geq 14$, then $9u^2 - 4u - 12 \geq 9u^2 - 6u + 2 \geq (3u - 1)^2 + 1$. Therefore, if $u \geq 7$, then $9u^2 - 4u - 12$ cannot be a perfect square. Checking through $u = 1, 2, \ldots, 6$, we find that $9u^2 - 4u - 12$ is only a perfect square for $u = 2$. Then by (1), $m = \frac{6 \pm \sqrt{36-20}}{2} = 1$ or 5. We see that $(2,1)$ and $(2,5)$ satisfy the desired property.

Note: This problem is materially equivalent to problem #4 from the 1994 International Math Olympiad [2].

12. *(2018 P6, Solved By: 0%, Average Score: 0, Author: Grant Molnar)*

We claim that the greatest common divisor (gcd) is always 1 (so 1 is possible, and it is the only possibility). The proof will use the fact that p is prime, but will actually be valid for any positive integer $q \geq 2$.

We employ a proof by contradiction: assume that the gcd is not 1; then there exists a prime number r such that $r \mid p-1$ and $r \mid \frac{q^p-1}{q-1}$. We divide our proof into cases: either $r \mid q - 1$, or $r \nmid q - 1$.

Case 1: $r \mid q - 1$.

In other words, $q \equiv 1 \pmod{r}$, so we have

$$0 \equiv \frac{q^p - 1}{q - 1} = \sum_{i=0}^{p-1} q^i \equiv \sum_{i=0}^{p-1} 1^i = p = (p-1) + 1 \equiv 1 \pmod{r},$$

since $r \mid p - 1$, which is a contradiction.

Case 2: $r \nmid q - 1$.

Then $\frac{q^p-1}{q-1} \equiv 0 \pmod{r}$ if and only if $q^p - 1 \equiv 0 \pmod{r}$; that is, we can ignore the denominator. We can rewrite this as $q^p \equiv 1 \pmod{r}$. Let a be the smallest positive integer such that $q^a \equiv 1 \pmod{r}$ (this is often called the *order of* $q \pmod{r}$). Then $q^p \equiv 1 \pmod{r}$ if and only if $a \mid p$.

On the other hand, Fermat's Little Theorem says that $q^{r-1} \equiv 1 \pmod{r}$, which implies $a \mid r - 1$. And since $r \mid p - 1$,

$$r - 1 < r \leq p - 1 < p,$$

so $a \mid r - 1$ implies $a < p$.

Therefore, $a \mid p$ and $a < p$. But p is prime, so $a = 1$. Finally, by definition of a, $q \equiv 1 \pmod{r}$, but that contradicts our assumption in Case 2 that $r \nmid q - 1$.

Note: This problem was first proven by the author in 2015 during his collaboration with the Brigham Young University Number Theory Seminar on the Feit-Thompson conjecture [19]. Their work, including this lemma, remains unpublished.

3.6 GEOMETRY

1. *(2015 Flyer, Author: Caleb Stanford)*

   (a) Starting at the midpoint of the hypotenuse of the first triangle, the distance to the midpoint of the hypotenuse of either of the new triangles is $\frac{1}{4}\sqrt{2}$. All future distances from this point are scaled by $\frac{1}{\sqrt{2}}$ at each iteration. Thus, the total distance traveled to the midpoint of *any* future hypotenuse drawn is bounded by the geometric series

   $$\frac{1}{4}\sqrt{2}\left[1+\left(\frac{1}{\sqrt{2}}\right)+\left(\frac{1}{\sqrt{2}}\right)+\left(\frac{1}{\sqrt{2}}\right)^2+\left(\frac{1}{\sqrt{2}}\right)^3+\cdots,\right]$$

   which converges because $\frac{1}{\sqrt{2}}<1$, and evaluates to

   $$\frac{\frac{1}{4}\sqrt{2}}{1-1/\sqrt{2}}=\frac{1/2}{\sqrt{2}-1}=\frac{1/2\cdot(\sqrt{2}+1)}{(\sqrt{2}-1)(\sqrt{2}+1)}=\frac{1}{2}+\frac{1}{2}\sqrt{2}.$$

   Now, any interior point of any triangle is always at most $\frac{1}{2}$ away from the midpoint of the hypotenuse, and hence at most

   $$\frac{1}{2}+\left(\frac{1}{2}+\frac{1}{2}\sqrt{2}\right)=1+\frac{1}{2}\sqrt{2}$$

   away from the origin of the plane. This means that the area is finite; indeed, it is bounded by πr^2 where r is the above radius, i.e.,

   $$\text{Area}\leq\pi\left(1+\frac{1}{2}\sqrt{2}\right)^2=9.15527\cdots.$$

   (b) We claim that the exact area is

   $$1+\frac{63}{64}=\frac{127}{64}=1.984375.$$

   The following diagrams show what is happening as the fractal figure is created.

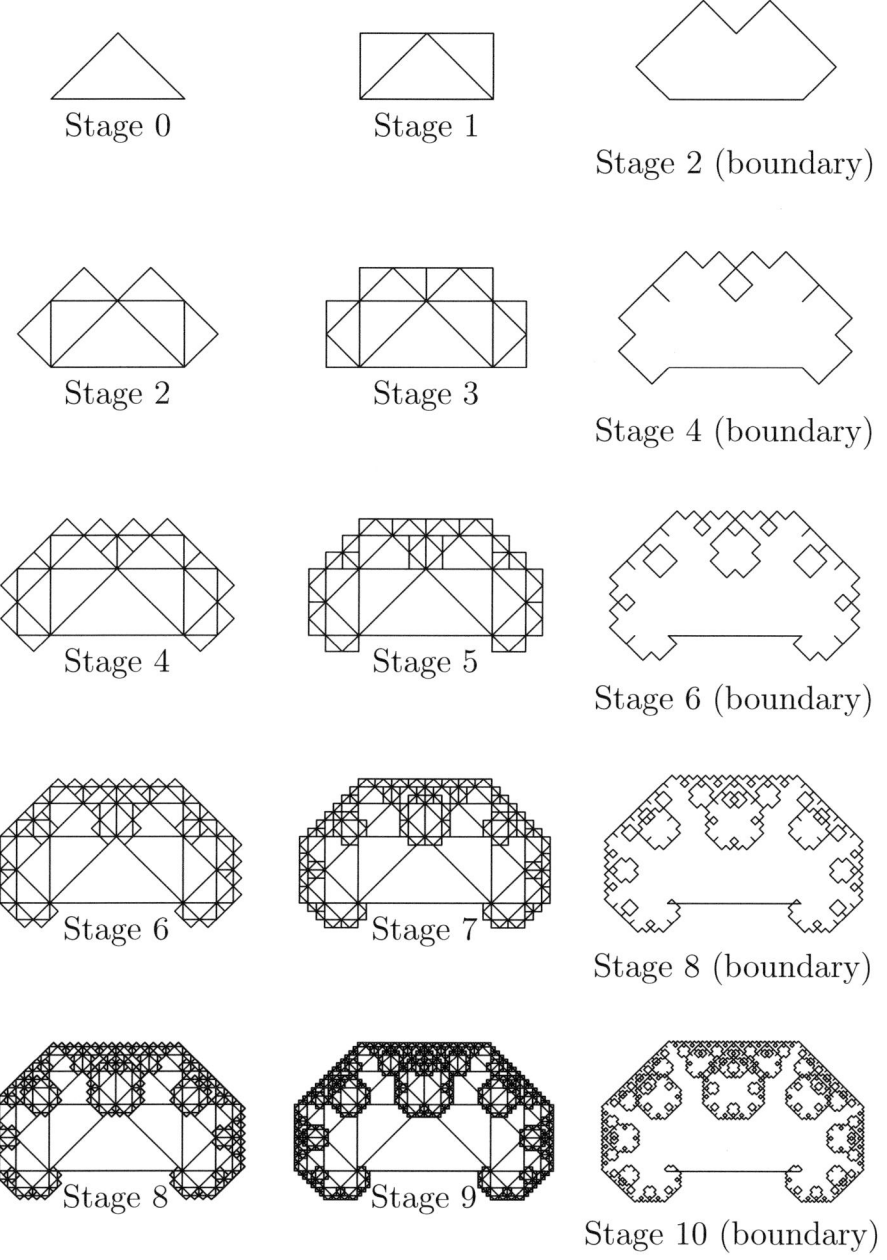

Stage 0

Stage 1

Stage 2 (boundary)

Stage 2

Stage 3

Stage 4 (boundary)

Stage 4

Stage 5

Stage 6 (boundary)

Stage 6

Stage 7

Stage 8 (boundary)

Stage 8

Stage 9

Stage 10 (boundary)

In the end, the fractal that is created has:

- A square of side length 1 in the center;
- To the left and to the right, a square of side length $\frac{1}{2}$ and a right triangle of side length $\frac{1}{2}$;
- Below each of these, a square of side length $\frac{1}{4}$ and a right triangle of side length $\frac{1}{4}$; and
- To the right and left of these, respectively, a square of side length $\frac{1}{8}$ and a right triangle of side length $\frac{1}{8}$, which wrap back around into the original square.

Before we prove this, we note that the answer follows from this analysis. In particular, using the fact that a square of side s has area s^2 and a triangle of side s has area $\frac{1}{2}s^2$, the total area is

$$1 + 2\left(\frac{1}{4} + \frac{1}{8}\right) + 2\left(\frac{1}{16} + \frac{1}{32}\right) + 2\left(\frac{1}{64} + \frac{1}{128}\right)$$
$$= 1 + \frac{32 + 16 + 8 + 4 + 2 + 1}{64}$$
$$= 1 + \frac{63}{64},$$

as claimed.

Now it remains to prove that the fractal has the regions that we claimed earlier. Let the midpoint of the hypotenuse of the first triangle be $(0,0)$ in the coordinate plane. We claim that it suffices to show that the region created:

(i) includes the entirety of the unit square with side $(-\frac{1}{2}, 0)$ to $(0, \frac{1}{2})$, above the y-axis; and

(ii) does not include any points above $y = 1$.

To see why this suffices, note that by the iterative process by which the fractal is created, rotating this initial region to the left and right and scaling, we get the left and right triangles, then the left and right unit squares, and so on. After Stage 7, the figure wraps around on itself and further iterations of conditions (i) and (ii) do not create any new information.

For both (i) and (ii), it is useful to think of the iterative process in sequences of two steps at a time. Stages 0 and 1 together form a rectangle; Stages 2 and 3 together created four rectangles from the first two rectangles, and so on.

- To show (i), note that after each two stages, we cover another rectangle with half the width of the previous one. Thus after stage 3, the diagram covers a rectangle with width 1 and height $\frac{1}{2} + \frac{1}{4}$, after stage 5 it covers a rectangle with width 1 and height $\frac{1}{2} + \frac{1}{4} + \frac{1}{8}$, and so on. Since $\frac{1}{2} + \frac{1}{4} + \frac{1}{8} + \cdots = 1$, in the limit it covers the unit square of height 1.

- To show (ii), we argue that after Stage $2n + 1$, the figure is entirely contained within a unit square centered at the origin of side length $1 + \frac{1}{2} + \frac{1}{4} + \cdots + \frac{1}{2^n}$. In the limit, it follows that the fractal is entirely contained within a unit square centered at the origin of side length 2, i.e., the figure is bounded by the lines $x = -1$ and $x = 1$ and by the lines $y = -1$ and $y = 1$. We can prove this by induction: note that after Stage $2n + 1$, the boundary of the unit square is formed by 2^n rectangles. The side length of the next rectangle created in two more stages is cut in half, while the number of rectangles doubles.

Overall, this shows that the figure is bounded inductively by the conditions (i) and (ii), and iterating these out one stage at a time, we get exactly the squares and triangles claimed as the boundaries to the figure. This completes the proof that the area is $1 + \frac{63}{64}$.

Note: Special thanks to a middle school student at MathPath, who came up with the fractal figure used in this problem.

2. *(2022 P1, Solved By: 71%, Average Score: 4.7, Author: Daniel South)*

The minimum n is 17.

Solution 1: Suppose that we place each square sequentially, so that the bottom left square is placed first. We place $n - 1$ additional squares, moving up and right by the same amount at each step. The final square is placed 1 unit up and 1 unit to the right of the initial square, so each new square is placed $\frac{1}{n-1}$ units up and $\frac{1}{n-1}$ units to the right of the previous square.

Therefore, when we place each new square, the area of overlap with the previous square is a square with side length $2 - \frac{1}{n-1}$, so the newly added area is

$$2^2 - \left(2 - \frac{1}{n-1}\right)^2 = \left(4 - \frac{1}{n-1}\right) \cdot \frac{1}{n-1}.$$

Since we add a total of $n - 1$ new squares, the total area of the region is

$$2^2 + (n-1) \cdot \left(\left(4 - \frac{1}{n-1}\right) \cdot \frac{1}{n-1}\right) = 8 - \frac{1}{n-1}.$$

Thus, we wish to find the smallest positive integer $n \geq 2$ such that $8 - \frac{1}{n-1} \geq \sqrt{63}$. This is equivalent to $8 - \sqrt{63} \geq \frac{1}{n-1}$, so $n - 1 \geq \frac{1}{8-\sqrt{63}}$. Multiplying by $\frac{8+\sqrt{63}}{8+\sqrt{63}}$, we find $n - 1 \geq 8 + \sqrt{63}$. Hence $n \geq 9 + \sqrt{63}$. The smallest such integer n is 17.

Solution 2: First, we extend the sides of the first and last squares, which shows that the region can be contained in a square of side length 3. Let $A = (-1, 1)$, $B = (0, 1)$, $C = (0, 2)$, and $D = (-1, 2)$. The top-left vertices of the 2×2 squares will all lie on $\overline{AC}$ as shown in the following picture.

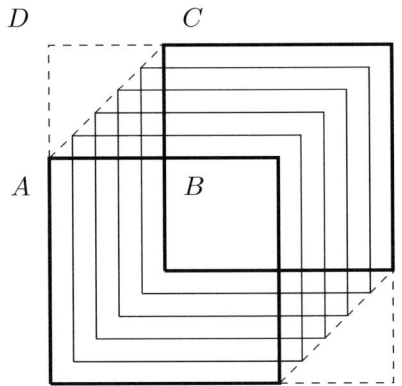

To find the area covered by the figure, we start with the 3×3 square with area $3 \cdot 3 = 9$. From this, we subtract the area of $\triangle ACD$ and its counterpart in the lower right-hand corner, obtaining an area of $9 - 2\left(\frac{1 \cdot 1}{2}\right) = 8$. In $\triangle ABC$, we must further subtract $n - 1$ small congruent isosceles right triangles. Since their leg lengths add up to 1, the leg length of each of these excluded triangles is $\frac{1}{n-1}$. The areas of these small triangles sums to $(n-1) \cdot \frac{(1/(n-1))^2}{2} = \frac{1}{2(n-1)}$. Since we must subtract the same amount in for the region in the lower right-hand corner, it follows that the total area covered by the paper is $8 - 2 \cdot \frac{1}{2(n-1)} = 8 - \frac{1}{n-1}$. As in Solution 1, the smallest integer n for which $8 - \frac{1}{n-1} > \sqrt{63}$ is 17.

3. *(2015 P2, Solved By: 45%, Average Score: 3.1, Author: Samuel Dittmer)*
The only possible values of AM are $\sqrt{697}\,/\,2$ and $\sqrt{1017}\,/\,2$.

First, let D be the foot of the altitude from A to $\overleftrightarrow{BC}$, so $AC = 13$, $AD = 12$, and $\angle ADC$ is a right angle. Then by the Pythagorean Theorem $CD = \sqrt{13^2 - 12^2} = 5$.

Now B must lie on $\overleftrightarrow{CD}$, but it must also lie on a circle of radius 20 centered at A. This gives two possible locations for the point B, which we draw below as B_1 and B_2, along with two possible locations for M, which we draw below as M_1 and M_2.

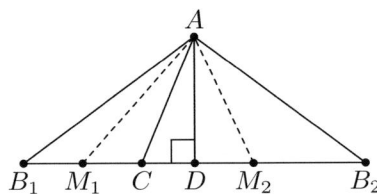

Here, we know that $B_1D = B_2D = \sqrt{20^2 - 12^2} = 16$. Therefore, $B_1C = 16 - 5 = 11$ and $B_2C = 16 + 5 = 21$. It follows that

$M_1C = 11/2$ and $M_2C = 21/2$. Hence $M_1D = 11/2 + 5 = 21/2$ and $M_2D = 21/2 - 5 = 11/2$. Thus $AM_1 = \sqrt{(21/2)^2 + 12^2} = \frac{\sqrt{1017}}{2}$ and $AM_2 = \sqrt{(11/2)^2 + 12^2} = \frac{\sqrt{697}}{2}$, and these are the two possible values of AM. These are clearly achievable as shown in the diagram above.

Note: Diagram due to Thomas Draper.

4. *(2017 P3, Solved By: 28%, Average Score: 2.7, Author: Peter Baratta)*

The desired ratio is $1 : 2$.

First we show that ABC is isosceles and $\angle ABC = 120°$. Since $\angle AOB$ and $\angle BOC$ are equal, chord lengths AB and BC are also equal. Also, $\triangle AOB$ and $\triangle BOC$ have equal sides $OA = OB = OC$ (the radius of the circle), and angles $\angle AOB = \angle BOC = 60°$, hence $\triangle AOB$ and $\triangle BOC$ are equilateral. Therefore, $\angle ABC = \angle ABO + \angle OBC = 60° + 60° = 120°$.

Let F be the foot of the altitude from C to $\overline{BP}$. Since FC is the shortest distance from C to the line through B and P, and B is on that line, we can deduce that $FC \leq BC$, with equality if and only if $F = B$.

Next we show that we can achieve $F = B$, so that $FC = BC$. Draw the line perpendicular to $\overline{BC}$ passing through B, and let P be the intersection of this line with $\overline{AC}$. Since $\angle ABC = 120°$, P lies between A and C. Choosing this P, we see that $\overline{BC}$ is perpendicular to $\overline{BP}$ by construction. Thus B is the foot of the altitude from C to $\overline{BP}$, so $F = B$. Thus we find that $FC = BC$ is achievable, so this P maximizes the length of the altitude from C to $\overline{BP}$. The location of P is unique because it is forced by $F = B$, and $F = B$ is the only way to maximize FC.

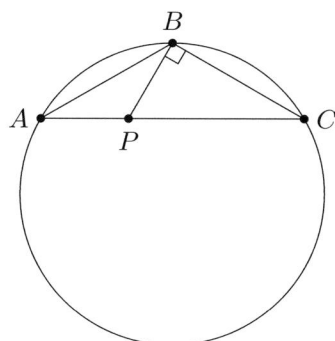

Without loss of generality let $AB = BC = 1$, so that $AC = \sqrt{3}$ (by the Law of Cosines, or splitting $\triangle ABC$ into two 30-60-90 triangles). Since $\triangle PBC$ is 30-60-90 with $BC = 1$, $PC = \frac{2}{3}\sqrt{3}$, which gives $AP = AC - PC = \frac{1}{3}\sqrt{3}$. Therefore, $\frac{AP}{PC} = \frac{1}{2}$ as desired.

5. *(2021 P2, Solved By: 14%, Average Score: 1.9, Author: Caleb Stanford)*

   The three circles C_1, C_2, and C_3 must have the same radius.

   Suppose that we only draw circles D and E, marking the points where C_1, C_2, and C_3 are tangent to E as T_1, T_2, and T_3, respectively. We also draw radii from the common center of D and E (call it O) to T_1, T_2, and T_3. Suppose that the radii intersect circle D at P_1, P_2, and P_3, respectively.

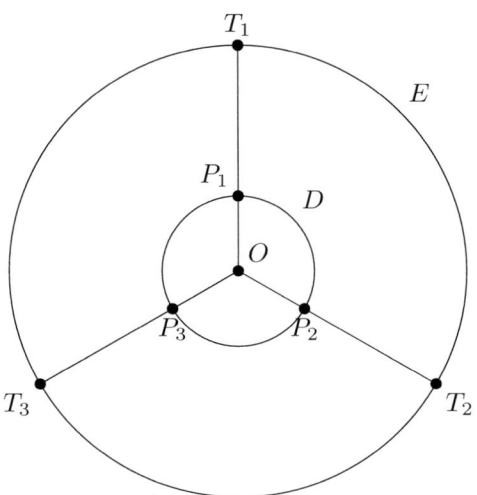

   If we were to draw the tangent lines to circle E at T_1, T_2, and T_3, then $\overline{OT_1}$, $\overline{OT_2}$, and $\overline{OT_3}$ would be perpendicular to the tangent line. The same applies for the radii of circles C_1, C_2, and C_3 drawn to T_1, T_2, and T_3, respectively. Hence the centers of C_1, C_2, and C_3 must lie on $\overline{OT_1}$, $\overline{OT_2}$, and $\overline{OT_3}$, respectively. Since the point of tangency between C_1 and D must lie on the line connecting the centers of the two circles, we see that P_1 must be the point of tangency between the two circles. It follows that $\overline{P_1T_1}$ is a diameter of circle C_1, and similarly, $\overline{P_2T_2}$ and $\overline{P_3T_3}$ are diameters of C_2 and C_3, respectively. If the radius of E is R and the radius of D is r, then $P_1T_1 = P_2T_2 = P_3T_3 = R-r$, so the three circles must have the same diameter, which implies that they must have the same radius.

   Note: Our diagram is not to scale—it turns out that circle D needs to be smaller. In fact, if the radius of circle D is 1, then C_1, C_2, and C_3 must have radius $3 + 2\sqrt{3}$, so E has radius $7 + 4\sqrt{3}$. The following picture shows a to-scale diagram.

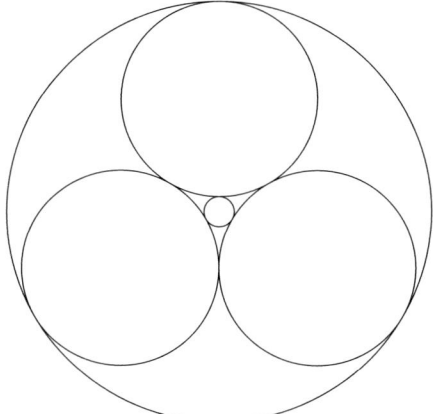

6. *(2018 P4, Solved By: 16%, Average Score: 1.5, Author: Caleb Stanford)*

Solution 1: When reflected across the sides of the square, the four triangles' boundaries line up. In particular, the triangles' boundaries which are *inside* the square lie on only 3 line segments: a line ℓ from vertex B, a line m from vertex C, and a line n from vertex D. (The reflections of BX and BY lie on ℓ, the reflections of CY and CZ lie on m, and the reflections of DZ and DX lie on n. Since X must be below and to the right of A, the two reflections of AX lie outside of the square.)

Orient the figure so that A is bottom-left, B top-left, C top-right, D bottom-right. Suppose toward a contradiction that there is a point not covered by the four reflected triangles. To not be covered by the reflection of XAB, it must be above ℓ. To not be covered by the reflection of YBC given that it is above ℓ, it must be below m. To not be covered by the reflection of ZCD given that it is below m, it must be below n. But if it is below n and inside the square then it is covered by the reflection of XDA. Contradiction, so no such point exists.

Solution 2: For the sake of contradiction, assume that there exists a point P inside of square $ABCD$ such that P is not covered by any of the reflections of the four triangles. For the moment, we will focus only on square $ABCD$. Let P', P'', P''', and P'''' be the reflections of P across lines AB, BC, CD, and DA, respectively, as shown in the following picture.

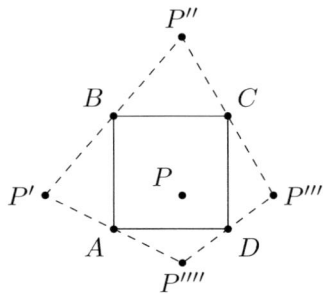

It can be seen that P', B, P'' are collinear, P'', C, P''' are collinear, P''', D, P'''' are collinear, and P'''', A, P' are collinear by comparing slopes. Now, since P is not covered by any of the four reflected triangles, P', P'', P''', and P'''' must all lie outside of triangle XYZ. Let r, s, and t be the lines extending the sides of triangle XYZ, such that r passes through B, s passes through C, and t passes through D. (In particular, X is the intersection of r and t, Y is the intersection of r and s, and Z is the intersection of s and t.)

Define "inside r" to mean below line r, "inside s" to mean below line s, and "inside t" to mean above line t. Then triangle XYZ consists of all points which are inside all of r, s, and t. Also, we may observe that P' is inside s and t, P'' is inside t, P''' is inside r, and P'''' is inside r and s. We summarize this with the following table:

	r	s	t
P'	?	inside	inside
P''	?	?	inside
P'''	inside	?	?
P''''	inside	inside	?

Because P' and P'' are collinear with B, which lies on r, one of them is outside r and one is inside r. Similarly, between P'' and P''', one is outside s and one is inside s. And between P''' and P'''', one is outside t and one is inside t. Subject to these constraints, it is impossible to fill out the above table without one row having *inside* in every column. This contradicts that P is not covered by any of the four reflections.

Note: This problem was adapted from the author's Mathematics Stack Exchange post [17].

7. *(2016 P4, Solved By: 17%, Average Score: 1.2, Author: Hiram Golze)*

We start by proving the following lemma about quadrilaterals that can have circles inscribed in them.

Lemma: *(Pitot's Theorem)* If quadrilateral $PQRS$ can have a circle inscribed inside of it, then $PQ + RS = QR + SP$. In other words, the sum of the opposite sides of such a quadrilateral are equal.

Proof: Given a quadrilateral $PQRS$ that can have a circle inscribed in it, let T, U, V, W be the points of tangency as shown below, and let O be the center of the circle.

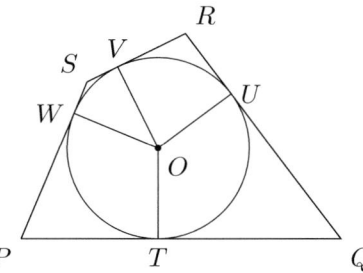

Then $\triangle OPT$ and $\triangle OPW$ are both right triangles with right angles at T and W, respectively. They also have $OP = OP$ and $OT = OW$, so by the Pythagorean Theorem, $WP = PT$. Similarly, $TQ = QU$, $UR = RV$, and $VS = SW$. In fact,

$$PQ + RS = PT + TQ + RV + VS = WP + QU + UR + SW = PS + RQ.$$

Thus in any such quadrilateral, the sum of opposite sides is equal. □

We will use this lemma. But we first derive some other facts. To start, we extend line AF through A and F, we extend line BC through B and C, and we extend line DE through D and E. If the extensions of $\overline{AF}$ and $\overline{BC}$ meet at K, the extensions of $\overline{BC}$ and $\overline{DE}$ meet at L, and the extensions of $\overline{DE}$ and $\overline{AF}$ meet at M, then $\triangle KLM$ must be equilateral. This follows because hexagon $ABCDEF$ is equiangular, so $\angle FAB = \angle ABC = 120°$, and therefore, $\angle BAK = 180 - \angle FAB = 60°$, $\angle KBA = 180 - \angle ABC = 60°$, so $\angle K = 180 - \angle BAK - \angle KBA = 60°$. Similar arguments show $\angle L = \angle M = 60°$. We draw the diagram below, where Z is the intersection of $\overline{AD}$ and $\overline{BE}$.

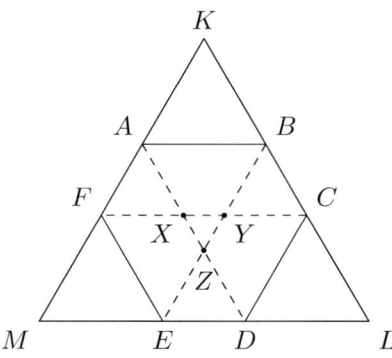

Note that $KB = AB$ and $CL = CD = AB$, hence $KL = 2AB + BC$. Similarly, $LM = 2AB + DE$ and $MK = 2AB + FA$. But triangle KLM is equilateral, so $KL = LM = MK$, and thus $BC = DE = FA$. Let $x = AB = CD = EF$ and let $y = BC = DE = FA$. Then as $AM = MD = x + y$, we know that $\triangle AMD \sim \triangle KML$, hence $\triangle AMD$ is equilateral. In particular, $\overline{AD} \parallel \overline{KL}$. By similar logic, $\overline{FE} \parallel \overline{KL}$. We can apply the same argument to find

$$\overline{AB} \parallel \overline{FC} \parallel \overline{ML} \text{ and } \overline{DC} \parallel \overline{EB} \parallel \overline{MK}.$$

This means that all of the angles in the above diagram are $60°$ or $120°$. It also implies that $ABYX$ is a trapezoid. It further implies $\triangle BAZ$ is equilateral, so $AZ = AB = x$. Also, $\triangle AFX$ is equilateral, so $AX = FA = y$.

As $x > y$, we find that $x = AZ \geq AX = y$, so X lies between A and Z. Therefore, $ZX = AZ - AX = x - y$. By symmetrical arguments, $XY = YZ = ZX = x - y$. As $\triangle BYC$ is equilateral, we also find $BY = BC = y$. Therefore, quadrilateral $ABYX$ is an isosceles trapezoid with $AB = x$, $BY = AX = y$, and $XY = x - y$. Applying the lemma, we find $AB + XY = BY + AX$, or $x + (x - y) = y + y$. Thus $2x = 3y$, and

$$\frac{AB}{FA} = \frac{x}{y} = \frac{3}{2}.$$

8. *(2014 P3, Solved By: 15%, Average Score: 1.1, Author: Hiram Golze)*

Let the legs of the right triangle be a and b, with $a \leq b$. Suppose that the right triangle satisfies the given properties. Then we have the following picture.

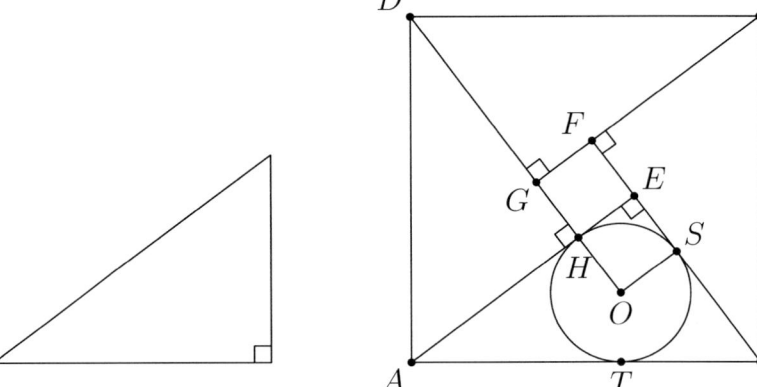

We know that $AE = b$, $BE = a$, and $AH = a$, therefore $EH = b - a$. We know that $EH = ES$, as they are tangents to circle O from the same point (or by power of a point), and hence $ES = b - a$. But as $BE = a$, this implies that $BS = BE - ES = a - (b - a) = 2a - b$. We also find that $TB = BS = 2a - b$, as they are tangents to circle O from the same point.

Similarly, as $AH = a$, we know that $AT = AH = a$. Therefore, $AB = AT + TB = a + (2a - b) = 3a - b$. Therefore, by the Pythagorean Theorem applied to triangle ABC, we know that

$$a^2 + b^2 = (3a - b)^2 = 9a^2 - 6ab + b^2$$

$$0 = 8a^2 - 6ab.$$

Therefore, $a(4a - 3b) = 0$. So either $a = 0$, which leads to degenerate triangles, or else $4a = 3b$. Therefore, we know that a is a multiple of 3, say $a = 3a'$ and b is a multiple of 4, say $b = 4b'$. Plugging this into the equation tells us that $12a' = 12b'$, or $a' = b'$. Therefore, all such triangles have legs of lengths $3n$ and $4n$, and a hypotenuse of length $5n$.

Now we must check that all such right triangles satisfy the desired property. Suppose we are given a right triangle ABE with side lengths $AB = 5n$, $BE = 3n$, and $EA = 4n$, where n is a positive integer, and points are labeled below.

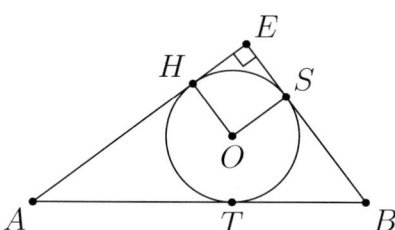

Then the perimeter is given by

$$12n = EH + HA + AT + TB + BS + SE = 2 \cdot EH + 2 \cdot AT + 2 \cdot TB$$

by the fact that $SE = EH$, $HA = AT$, and $TB = BS$. Therefore,

$$2(EH + AT + TB) = 2(EH + AB) = 2(EH + 5n) = 12n,$$

so we conclude that $EH = SE = n$. We can use a similar method to show that $HA = AT = 3n$ and $TB = BS = 2n$. Therefore, $EB = HA$. Now we arrange two of the triangles into a part of the square formation.

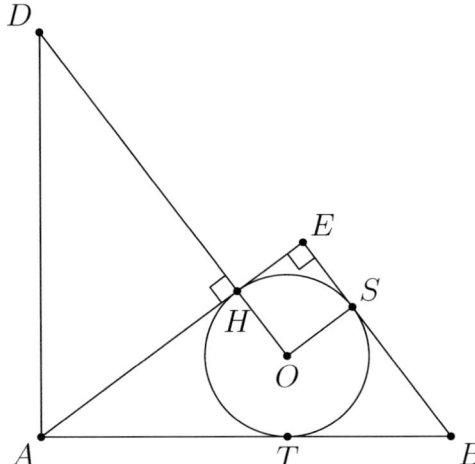

As $AH = 3n$, angle AHD is right, and $\overline{OH} \perp \overline{AE}$, we know that D, H, and O are collinear. Because $\overline{DH}$ contains the side of a smaller square, we know that O lies on the extension of the smaller square. By symmetry, the same must apply for the other three triangles. Therefore, all triangles with sides lengths $3n$, $4n$, and $5n$, where n is a positive integer do indeed satisfy the desired property. Thus they describe all such triangles.

9. *(2020 P4, Solved By: 11%, Average Score: 0.8, Author: Josh Speckman)*

   Solution 1: Let O be the center of C. First, we draw a diagram:

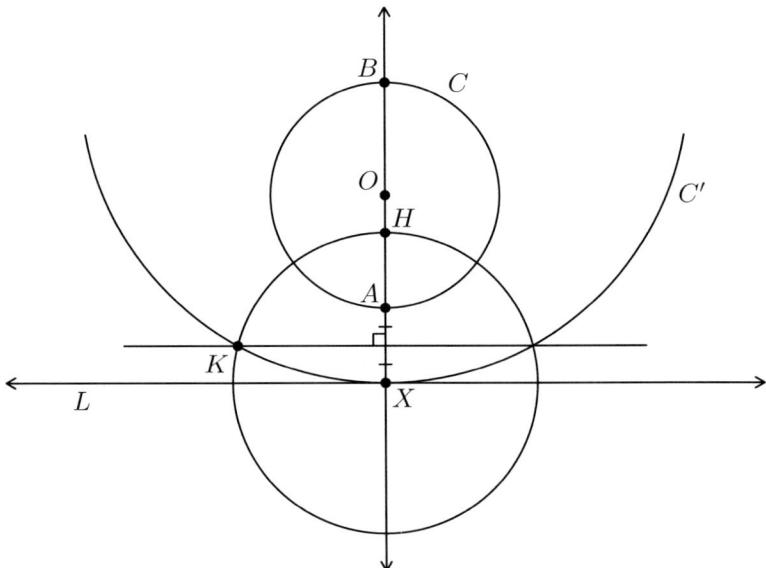

Note that X, A, O, B are collinear (and lie in that order). Let $r = AO = OB$ be the radius of C, and let $XA = 2a$.

We first calculate the length of $\overline{XK}$. Let K' be the midpoint of $\overline{AX}$. By Pythagorean Theorem,

$$XK^2 = K'K^2 + K'X^2 = K'K^2 + a^2.$$

Also by Pythagorean Theorem,

$$K'K^2 = KB^2 - K'B^2 = XB^2 - K'B^2$$
$$= (2r + 2a)^2 - (2r + a)^2 = 4ar + 3a^2.$$

Combining these,

$$XK^2 = (4ar + 3a^2) + a^2 = 4ar + 4a^2$$
$$\implies \quad XK = 2\sqrt{a^2 + ar}.$$

And $XH = XK$ by definition, so

$$XH = 2\sqrt{a^2 + ar}, \tag{1}$$

which is between $2a$ and $2a + r$. This confirms that H lies between A and O, as pictured in the diagram.

Let P be any point on line L. Let $x = XP$, and let C_P be the circle centered at P passing through H.

The following is a standard fact about perpendicular circles: *Two circles are perpendicular if and only if $r_1^2 + r_2^2 = d^2$, where r_1, r_2 are the radii of the circles and d is the distance between the two centers.*

Using this fact, we will show that C_P is perpendicular to C. We need to calculate the radii of the two circles and the distance between the centers. First, the radius of C is

$$r_1 = r. \tag{2}$$

Second, the radius of C_P is PH, and by Pythagorean Theorem this is

$$r_2^2 = XH^2 = XP^2 + XH^2 = x^2 + 4a^2 + 4ar, \tag{3}$$

by (1). Third, the distance d is

$$d^2 = PO^2 = XP^2 + XO^2 = x^2 + (2a + r)^2 = x^2 + 4a^2 + 4ar + r^2. \tag{4}$$

Now we can see that adding (2) and (3), we get (4), so $r_1^2 + r_2^2 = d^2$, and we are done.

Solution 2: We solve this using coordinates. Let $X = (0,0)$, $A = (0, a)$, and $B = (0, b)$, so line L has equation $y = 0$.

If O is the center of circle C, then $O = (0, \frac{a+b}{2})$. We know K has y-coordinate $\frac{a}{2}$, so the difference between the y-coordinates of B and K is $b - a/2$.

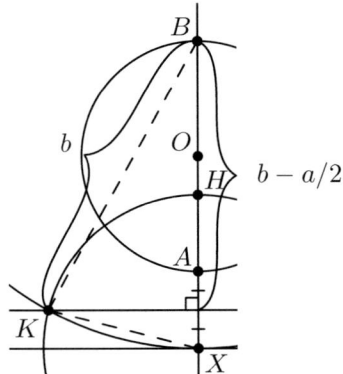

Since K is on the circle of radius b centered at B, by the Pythagorean Theorem, it has x-coordinate $-\sqrt{b^2 - (b - a/2)^2} = -\sqrt{ab - a^2/4}$. Therefore, as $K = (-\sqrt{ab - a^2/4}, a/2)$ and $X = (0, 0)$,

$$KX = \sqrt{(ab - a^2/4) + a^2/4} = \sqrt{ab}.$$

It follows that $H = (0, \sqrt{ab})$.

Now let $P = (r, 0)$, and let $Q = (x, y)$ be a point where the circle through H centered at P intersects C. Then since Q lies on both circles, it must satisfy

$$x^2 + \left(y - \frac{a + b}{2}\right)^2 = \left(\frac{a - b}{2}\right)^2$$
$$(x - r)^2 + y^2 = PH^2 = ab + r^2.$$

Adding these equations, we find

$$2x^2 - 2rx + r^2 + 2y^2 - (a + b)y + (a + b)^2/4 = ab + (a - b)^2/4 + r^2.$$

This simplifies to

$$x^2 + y^2 - rx - (a + b)y/2 = 0. \tag{1}$$

On the other hand, note that the dot product of $\overrightarrow{PQ}$ and $\overrightarrow{OQ}$ is

$$\overrightarrow{PQ} \cdot \overrightarrow{OQ} = (x - r, y) \cdot \left(x, y - \frac{a + b}{2}\right)$$
$$= x^2 - rx + y^2 - (a + b)y/2$$
$$= 0,$$

where in the last step, we used (1). This implies that $\overrightarrow{PQ}$ and $\overrightarrow{OQ}$ are perpendicular, so the circles themselves must be perpendicular.

10. *(2013 P4, Solved By: 0%, Average Score: 0.5, Author: Hiram Golze)*

We claim that it is necessary and sufficient that $\beta + \gamma = \frac{\pi}{2}$.

To prove that this condition is necessary, we do some angle chasing. Suppose that such a triangle XYZ exists with Z on L_3 as shown.

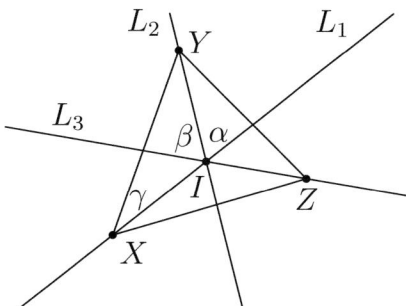

Now $\angle YIX = \pi - \alpha$, so by triangle XYI, we know that $\angle XYI = \pi - (\pi - \alpha) - \gamma = \alpha - \gamma$. By our assumption that the lines are angle bisectors, we have $\angle IYZ = \alpha - \gamma$ as well. Similarly, $\angle IXZ = \gamma$. Now $\angle YIZ = \pi - \beta$, so using triangle IYZ, we have $\angle YZI = \pi - (\pi - \beta) - (\alpha - \gamma) = \beta - \alpha + \gamma$. Using vertical angles, we know that $\angle XIZ = \alpha + \beta$. Thus by triangle XIZ, we know that $\angle IZX = \pi - (\alpha + \beta) - (\gamma) = \pi - \alpha - \beta - \gamma$. As L_3 bisects $\angle XZY$, we know that $\angle XZI = \angle YZI$, or rather $\pi - \alpha - \beta - \gamma = \beta - \alpha + \gamma$. Simplifying this, we get $\frac{\pi}{2} = \beta + \gamma$. Therefore, for such a triangle to exist, it is necessary that $\beta + \gamma = \frac{\pi}{2}$.

We claim that this condition is also sufficient. Suppose $\beta + \gamma = \frac{\pi}{2}$. First, let's ignore L_3 for the moment as shown below, to the left. From points X and Y, we draw lines at angles γ and $\alpha - \gamma$, respectively. If we call their intersection point Z, then L_1 and L_2 are angle bisectors of $\triangle XYZ$ as shown below, to the right. Then the line passing through I and Z must also be an angle bisector of $\triangle XYZ$ by concurrency of angle bisectors. We claim that line IZ is in fact line L_3, which would tell us that this condition is sufficient.

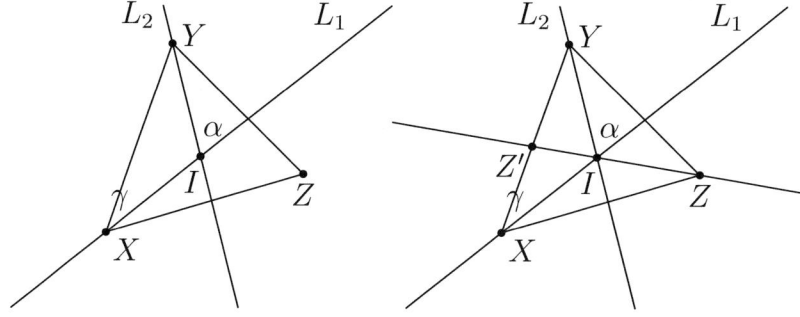

We proceed again by angle chasing. Once again, $\angle YIX = \pi - \alpha$, so by triangle XYI, we know that $\angle XYI = \pi - (\pi - \alpha) - \gamma = \alpha - \gamma$. As the three lines are all angle bisectors of triangle XYZ, we know that $\angle YZX = \pi - 2(\gamma) - 2(\alpha - \gamma) = \pi - 2\alpha$. Thus $\angle IZY = \angle IZX = \frac{\pi}{2} - \alpha$. By triangle IYZ, we have $\angle YIZ = \pi - (\alpha - \gamma) - (\frac{\pi}{2} - \alpha) = \frac{\pi}{2} + \gamma$. If Z' is the intersection of line ZI with line XY, then we know that $\angle YIZ' = \pi - (\frac{\pi}{2} + \gamma) = \frac{\pi}{2} - \gamma$. By our assumption that $\beta + \gamma = \frac{\pi}{2}$, we know that this angle is just β. But line L_3 was drawn at angle β from L_2, so line L_2 is in fact our phantom line ZI. Therefore, triangle XYZ is the desired triangle, so this condition is sufficient.

11. *(2015 P6, Solved By: 0%, Average Score: 0.3, Author: Hiram Golze)*

First, suppose that the plane that contains the upper circular base of the cylinder intersects $\overline{OA}$, $\overline{OB}$, and $\overline{OC}$ in the points A', B', and C', respectively. We claim that if $OABC$ achieves the minimum possible height, then the upper circular base of the cylinder is the inscribed circle of $\triangle A'B'C'$.

If not, then we may suppose without loss of generality that $\overline{A'B'}$ is not tangent to the circular base. Then, within the triangle $\triangle A'B'C'$ we can draw line $A''B''$ parallel to side $\overline{A'B'}$, such that A'' is on $\overline{A'C'}$ and B'' is on line $B'C'$ and $\overline{A''B''}$ is tangent to the circular base. As $\overline{A''B''}$ is parallel to $\overline{A'B'}$, which in turn is parallel to $\overline{AB}$, then A, B, A'', B'' are coplanar.

Let O' be the intersection of plane $ABA''B''$ with line OC. The new pyramid $O'ABC$ contains the cylinder, and it has strictly smaller height than pyramid $OABC$. It also has a higher number of faces tangent to the upper circular base of the cylinder than pyramid $OABC$. By applying this process on each side of the triangle ABC, we obtain a pyramid with strictly smaller height and all lateral faces tangent to the upper circular base. Therefore, given a triangular base ABC and apex O, we can always find a pyramid with smaller height and all three lateral faces tangent to the upper circular base of the cylinder. Hence we may assume that the upper circular base of the cylinder is the incircle (inscribed circle) of $\triangle A'B'C'$.

Now as face ABC is parallel to face $A'B'C'$, we know that pyramids $OABC$ and $OA'B'C'$ are similar. Therefore, the similarity ratio tells us that if r is the inradius (radius of the incircle) of $\triangle ABC$ and if h is the height from O of pyramid $OABC$, then

$$\frac{\text{inradius}(ABC)}{\text{inradius}(A'B'C')} = \frac{\text{height}(OABC)}{\text{height}(OA'B'C')}$$

$$\frac{r}{4} = \frac{h}{h - 10}.$$

Solving for h, we find that

$$h = \frac{10r}{r - 4} = 10 + \frac{40}{r - 4}.$$

Clearly, $r > 4$, and by the above equation, as r increases, the value of h must decrease. So the problem simplifies to finding the maximum possible inradius of a triangle with perimeter 84.

In $\triangle ABC$, let D, E, and F be the points of tangency of the incircle with sides $\overline{BC}$, $\overline{CA}$, and $\overline{AB}$, respectively. Also, let $x = AE = AF$, $y = BD = BF$, $z = CD = CE$. Then the perimeter of $\triangle ABC$ is

$$2(x + y + z) = 84.$$

Hence the semiperimeter s (i.e., half the perimeter) is given by $s = x + y + z = 42$. Note that $x = s - a$, $y = s - b$, and $z = s - c$. Also, the area of $\triangle ABC$ (by the inradius formula and Heron's formula) is

$$42r = \sqrt{42xyz}.$$

Hence

$$r = \sqrt{\frac{xyz}{42}}.$$

By the AM-GM inequality,

$$xyz \leq \frac{(x + y + z)^3}{27} = 14^3,$$

with equality if and only if $x = y = z$, i.e., $\triangle ABC$ is equilateral. Hence

$$r = \sqrt{\frac{xyz}{42}} \leq \frac{14}{\sqrt{3}}.$$

This leads to a height of

$$h = \frac{490 + 140\sqrt{3}}{37},$$

achieved when we have base ABC with side lengths $AB = BC = CA = 28$.

12. *(2019 P5, Solved By: 0%, Average Score: 0.1, Author: Samuel Dittmer)*

Angle C must be 90 degrees.

Suppose that x, y, and z are the lengths of the tangents from A, B, and C, respectively to the incircle.

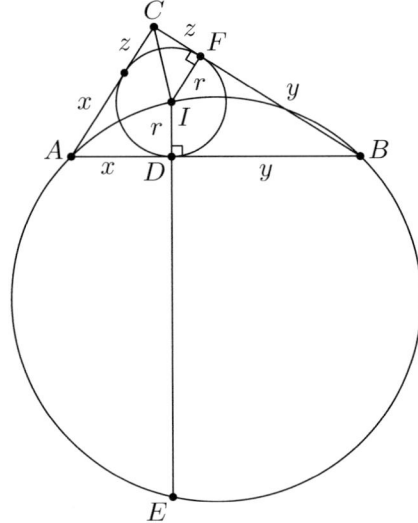

First we derive a formula for the inradius r of triangle ABC in terms of $x, y,$ and z, and the semiperimeter $s = \frac{a+b+c}{2}$. Note that $a = y + z$, $b = x + z$, and $c = x + y$. Adding these, we find $a + b + c = 2(x + y + x)$, hence $s = x + y + z$. Subtracting $a = y + z$, $b = x + z$, and $c = y + z$, respectively, from $s = x + y + z$, we find

$$s - a = x$$
$$s - b = y$$
$$s - c = z.$$

Therefore, substituting these into Heron's formula, we find $[ABC] = \sqrt{sxyz}$. Since the area of any triangle is equal to its inradius times its semiperimeter, we also have $[ABC] = rs$ (where r is the inradius). Thus $rs = \sqrt{sxyz}$, and

$$r = \sqrt{\frac{xyz}{s}}. \tag{1}$$

Next, we consider $\angle C$. We know that I lies on the angle bisector of $\angle C$, so if F is the foot of the perpendicular from I to $\overline{BC}$, then $\triangle IFC$ is right with $\angle ICF = \frac{\angle C}{2}$. Therefore,

$$\tan \frac{\angle C}{2} = \tan \angle ICF = \frac{r}{z},$$

and applying the formula for r in (1),

$$\tan \frac{\angle C}{2} = \frac{\sqrt{xyz/s}}{z} = \sqrt{\frac{xy}{sz}}. \tag{2}$$

Finally, we apply power of a point to point D and the circumcircle of $\triangle ABI$. We find $BD \cdot AD = ED \cdot DI$. Therefore, since $BD = y$, $AD = x$, $DI = r$, and $ED = IE - DI = b + a - r$, we find

$$xy = (b + a - r) \cdot r. \tag{3}$$

Note that $b + a = (x + z) + (y + z) = s + z$. Therefore, replacing $b + a$ by $s + z$ in (3), and replacing r using equation (1), we find

$$xy = \left((s + z) - \sqrt{\frac{xyz}{s}} \right) \cdot \sqrt{\frac{xyz}{s}}.$$

Hence $xy = (s + z)\sqrt{\frac{xyz}{s}} - \frac{xyz}{s}$, or

$$xy + \frac{xyz}{s} = (s + z)\sqrt{\frac{xyz}{s}}.$$

We can factor this as

$$xy \left(1 + \frac{z}{s} \right) = (s + z)\sqrt{\frac{xyz}{s}}.$$

Therefore,

$$xy \left(\frac{s + z}{s} \right) = (s + z)\sqrt{\frac{xyz}{s}}.$$

Dividing by all the terms on the right-hand side, we find

$$\sqrt{\frac{xy}{sz}} = 1.$$

Therefore, by (2), $\tan \frac{\angle C}{2} = 1$. Hence $\frac{\angle C}{2} = 45°$, and $\angle C = 90°$.

Bibliography

[1] Gary Antonick. The Cupcake Puzzle — Numberplay. *The New York Times*, 2013. https://nyti.ms/443beI9 [Accessed January 2024].

[2] IMO Board. 35th IMO 1994 — International Mathematical Olympiad. https://www.imo-official.org/year_info.aspx?year=1994, 1994. [Accessed January 2024].

[3] Marijke Hans L Bodlaender, Cor AJ Hurkens, Vincent JJ Kusters, Frank Staals, Gerhard J Woeginger, and Hans Zantema. Cinderella versus the wicked stepmother. In *7th International Conference on Theoretical Computer Science (TCS)*, pages 57–71. Springer, 2012.

[4] Alexandru Buium. *Arithmetic Differential Equations*. Number 118 in Mathematical surveys and monographs. American Mathematical Society, 2005.

[5] William Gasarch, Erik Metz, Jacob Prinz, and Daniel Smolyak. *Mathematical Muffin Morsels: Nobody wants a small piece*, volume 16. World Scientific, 2020.

[6] Owen Healy. The Rectangle Puzzle — Puzzling Stack Exchange. https://puzzling.stackexchange.com/q/40781/203, 2016. [Accessed January 2024].

[7] Antonius JC Hurkens, Cor AJ Hurkens, and Gerhard J Woeginger. How Cinderella won the bucket game (and lived happily ever after). *Mathematics Magazine*, 84(4):278–283, 2011.

[8] OEIS Foundation Inc. Sequence A002454: Central factorial numbers: $a(n) = 4^n (n!)^2$. — the On-line Encyclopedia of Integer Sequences. https://oeis.org/A002454, 1991. [Accessed January 2024].

[9] OEIS Foundation Inc. Sequence A056570: Third power of Fibonacci numbers. — the On-line Encyclopedia of Integer Sequences. https://oeis.org/A056570, 2000. [Accessed January 2024].

[10] OEIS Foundation Inc. Sequence A269869: Number of matchings (not necessarily perfect) in the triangle graph of order n. — the On-line Encyclopedia of Integer Sequences. https://oeis.org/A269869, 2016. [Accessed January 2024].

[11] OEIS Foundation Inc. Sequence A359576: Array read by antidiagonals: $T(m, n)$ is the number of $m \times n$ binary arrays with a path of adjacent 1's from top row to bottom row. — the On-line Encyclopedia of Integer Sequences. https://oeis.org/A359576, 2023. [Accessed January 2024].

[12] OEIS Foundation Inc. Sequence A362014: Number of distinct lines passing through exactly two points in a triangular grid of side n. — the On-line Encyclopedia of Integer Sequences. https://oeis.org/A362014, 2023. [Accessed January 2024].

[13] OEIS Foundation Inc. Sequence A365988: Number of $n \times n$ binary arrays with a path of adjacent 1's from top row to bottom row. — the On-line Encyclopedia of Integer Sequences. https://oeis.org/A365988, 2023. [Accessed January 2024].

[14] Bennet Manvel. Counterfeit coin problems. *Mathematics Magazine*, 50(2):90–92, 1977.

[15] Cedric AB Smith. The counterfeit coin problem. *The Mathematical Gazette*, pages 31–39, 1947.

[16] Canadian Mathematical Society. CMO 1975 — Canadian Mathematical Olympiad. https://cms.math.ca/wp-content/uploads/2019/07/exam1975.pdf, 1975. [Accessed January 2024].

[17] Caleb Stanford. Unit square inside triangle (answer) — Mathematics Stack Exchange. https://math.stackexchange.com/q/1953443/, 2017. [Accessed January 2024].

[18] Caleb Stanford. Who wins the election between the positive integers 1 through 10? – Mathematics Stack Exchange. https://math.stackexchange.com/q/3591160, 2020. [Accessed January 2024].

[19] NM Stephens. On the Feit-Thompson conjecture. *Mathematics of Computation*, 25(115):625, 1971.

[20] Harvard/MIT Math Tournament. HMMT 2003 — Harvard/MIT Math Tournament. https://www.hmmt.org/www/archive/62, 2013. [Accessed January 2024].

[21] Peter Winkler. *Mathematical puzzles*. AK Peters/CRC Press, 2020. ISBN-13 978-0367206925.

[22] Paul Zeitz. *The art and craft of problem solving*. John Wiley & Sons, 2017.

Index

AM-GM inequality, 75, 113
approximation, 50, 51
arithmetic differential equations, 76
asymptotic growth, 8, 50, 52, 67, 81
author
>   Annie Yun, 68
>   Benjamin Lovelady (solution),
>   72
>   Benjamin Stanford, 78
>   Caleb Stanford, 23–25, 28, 29,
>   31, 34, 40, 43, 49, 51, 65,
>   67, 68, 70, 72, 74, 84, 91,
>   92, 96, 102, 103
>   Daniel South, 47, 75, 79, 99
>   Grant Molnar, 23, 33, 35, 38, 54,
>   61, 64, 73–77, 82, 95
>   Hiram Golze, 27, 30, 36, 41, 57,
>   62, 66, 74, 81, 83, 85, 86,
>   89, 104, 106, 111, 112
>   Josh Speckman, 108
>   Multiple, 74, 83
>   Peter Baratta, 101
>   Samuel Dittmer, 59, 71, 80, 81,
>   83, 87, 100, 113
>   Thomas Draper (diagram), 101
>   Unknown, 81
>   Wyatt Mackey, 83

bijection proofs, 40, 42, 43, 48, 54,
    57
binomial coefficient, 18, 24, 33, 34,
    36–38, 51–53, 55, 81, 82
Binomial Theorem, 81

Chinese Remainder Theorem, 88, 89
coloring, 7, 8, 10, 11, 29, 34, 43, 45,
    50
coloring argument, 28, 29
counterfeit coin problem, 5, 24

cupcake puzzle, *see* muffin problem

Diophantine equations, 17–19
discrete dynamical systems, 7, 8

Euclidean algorithm, 83, 86
Euler characteristic, 36
expected value, 9, 36, 38

factorial numbers, 18, 19, 38, 39, 47,
    73, 84, 85, 91
>   central, 43
>   falling, 43
Feit-Thompson conjecture, 95
Fermat's Little Theorem, 95
Fibonacci numbers, 57
floor function, 52, 68, 92
flyer
>   problems, 5, 8, 11, 12, 16–19
>   solutions, 23, 33, 59, 61, 72–74,
>   81, 96
fractal, 19, 96–99
functional equations, 17, 18

game equivalence argument, 59, 68,
    70, 71
geometric series, 92, 96
graph theory, 6, 8, 9, 13, 27, 28, 34,
    36
greatest common divisor, 18, 19, 31,
    82, 83, 86, 87, 95

Heron's formula, 113, 114

incircles and circumcircles, 21, 22,
    112–115
inclusion-exclusion, 33, 36, 39, 50, 51
indefinite play, *see* infinite play
induction, 33, 48, 49, 57, 68, 73, 76,
    99

backward, 70
strong, 56
infinite play, 15, 67, 68
inscribed circle, 21
invariant, 31

Law of Cosines, 86, 87, 101
linearity of expectation, 36

method of finite differences, 90
misère rule of play, 12
modular arithmetic, 11, 19, 31,
        47–49, 52, 54, 55, 75, 82,
        84, 85, 87, 89, 90, 95
muffin problem, 16, 73
multiplayer games, 14, 15

necessary and sufficient conditions,
        21, 111
not to scale, 102
number of divisors function, 19, 91

one-to-one correspondence, *see*
        bijection proofs
open questions, 51, 59, 71, 95
optimization problems, 6, 16, 17, 22,
        52

parity argument, 29, 75, 82, 83,
        85–87, 90, 91
partitions, 10, 43, 46
Pascal's triangle, 82, *see also*
        binomial coefficient
path (in a graph), 7, 9, 27–29, 41, 61
Pell's equation, 87
perpendicular circles, 21, 109
pigeonhole principle, 24, 73
Pitot's Theorem, 105
polynomials, 16, 17, 19, 53, 72,
        74–76, 82, 83, 89, 90, 92
discriminant of, 93
Power of a Point, 107, 115
probability, 9, 11, 38–40, 52
problems
        2013, 6, 11, 15, 16, 18, 19, 21
        2014, 7, 9, 12, 16, 18, 21

        2015, 9, 17–20, 22
        2016, 7, 9, 11, 13, 18, 19, 21
        2017, 5, 6, 8, 11, 15, 19, 20
        2018, 6, 8, 10, 13, 17, 19, 20
        2019, 5, 8, 11, 14, 17, 18, 22
        2020, 6, 9, 15, 17, 18, 21
        2021, 10, 12, 15, 17, 19, 20
        2022, 7, 11, 15, 17, 18, 20
Pythagorean Theorem, 100, 105, 107,
        109, 110
Pythagorean triples, 86

quadratic formula, 93

recurrence relation, 17, 48, 49, 55,
        57, 73, 74
homogeneous linear, 49
reflection (geometry), 20, 103
relatively prime, 12, 18, 31, 32, 59,
        82

sequences, 10, 11, 17, 18, 48, 49, 55,
        76, 80
Online Encyclopedia of Integer
        Sequences (OEIS), 34, 43,
        51, 59
solutions
        2013, 27, 57, 66, 71, 81, 87, 111
        2014, 30, 41, 62, 72, 81, 85, 106
        2015, 40, 78, 81, 83, 96, 100, 112
        2016, 29, 38, 59, 64, 80, 92, 104
        2017, 23, 24, 34, 54, 67, 89, 101
        2018, 25, 33, 43, 64, 74, 95, 103
        2019, 23, 35, 51, 65, 75, 84, 113
        2020, 28, 36, 70, 73, 79, 86, 108
        2021, 47, 61, 68, 75, 77, 91, 102
        2022, 31, 49, 68, 74, 76, 82, 99
strategy stealing, 62
symmetry argument, 35, 39, 62, 106,
        108
systems of equations, 17, 18, 74, 81

tangent
        circles, 20, 102
        lines, 21, 22, 102, 106, 112, 113
        planes, 112

tiling, 5, 8, 11, 13, 23, 35, 57, 64
triangular numbers, *see* binomial
      coefficient
Vieta's formulas, 72, 74

weighted mean, 67
wicked stepmother problem, 68
winning strategy construction, 23,
      59, 62, 64, 65, 67, 69, 72